MINI
WEAPONS 3
OF MASS DESTRUCTION
BUILD SIEGE WEAPONS OF THE DARK AGES

JOHN AUSTIN

CHICAGO
REVIEW
PRESS

Library of Congress Cataloging-in-Publication Data

Austin, John, 1978–

  Miniweapons of mass destruction. 3 : build siege weapons of the Dark Ages / John Austin.

   pages cm

  ISBN 978-1-61374-548-9

  1. Siege warfare–Miscellanea. 2. Miniature weapons. 3. Handicraft. 4. War toys. I. Title.
II. Title: Build siege weapons of the Dark Ages.

  UG444.A92 2013

  623.4'41–dc23

                        2013002774

Cover and interior design: Jonathan Hahn

Illustrations: Austin Design, Inc.

© 2013 by Austin Design, Inc.

All rights reserved

Published by Chicago Review Press, Incorporated

814 North Franklin Street

Chicago, Illinois 60610

ISBN 978-1-61374-548-9

Printed in the United States of America

10 9 8 7 6 5 4 3 2 1

This book is lovingly dedicated to William:
may you conquer any obstacle in your path.
Love always, Dad

To the knight-in-training who wields this weapon of
knowledge, that you solemnly swear an oath of allegiance
to the throne, to use the contents of this book only for
the greater good of the kingdom, and to refrain from the
unjust use of MiniWeapons against your fellow man. If
you honor this agreement, you will be recognized with a
swift promotion from page to squire to knighthood.

Join the MiniWeapons army on Facebook:
MiniWeapons of Mass Destruction: Homemade Weapons Page

For video demos, tutorials, and other extras, find us on YouTube:
MiniWeaponsBook Channel

CONTENTS

Introduction   vii
Play It Safe   ix

1 Rubber Band Catapults   **1**
   Tic Tac Catapult   3
   Card Box Catapult   9
   Ruler Catapult   19
   Candy Box Catapult   25
   Pen and Pencil Catapult   33
   Metal Hanger Catapult   39
   Armored Catapult   49

2 Nonelastic Catapults   **57**
   Gift Card Catapult   59
   Tic Tac Onager   65
   Depressor Spoon Catapult   71
   Marshmallow Catapult   75
   Mousetrap Catapult   79
   CD Trebuchet   87

3 Bows and Arrows   **101**
   Double Skewer Bow   103
   Chopstick Bow   107

Plasticware Bow 111

Plastic Hanger Bow 119

Advanced Pen Bow 127

Composite Ruler Bow 135

4 Crossbows   141

Wooden Ruler Crossbow 143

Plasticware Crossbow 149

Craft Stick Crossbow 157

Plastic Ruler Crossbow 163

Office Crossbow 169

Compound Crossbow 175

5 Ballistae   183

Gift Card Ballista 185

Pen Ballista 191

Clothespin Ballista 201

Mounted Siege Ballista 213

6 Mini Siege   225

Craft Stick Mini Bow 227

Paper Clip Bow 231

Tri-Clip Bow 235

Bottle Cap Crossbow 239

Toothpick Crossbow 243

Clip and Cap Catapult 251

7 Targets   255

Carton Siege Tower 257

Attacking Army 259

Tower of Oatmeal 261

Storm the Castle 263

Print-Out Targets 265

INTRODUCTION

Fill your armory with *MiniWeapon of Mass Destruction 3*, a homemade weapons guide that will help you transform everyday items into catapults, crossbows, and other instruments of siege warfare.

This collection of MiniWeapons shouldn't be taken lightly. Inspired by the battlefields of the 5th to the 15th century, this stockpile of medieval marvels has a single purpose: to bombard the enemy with artillery, whether to blockade reinforcements, force armies to surrender, or crumble fortified castles.

The projects in this book can be built quickly, giving you a definite tactical advantage—the element of surprise! Each project comes with a list of easy-to-obtain materials and illustrated step-by-step instructions. And in the final chapter, you'll find a collection of simple targets direct from the Dark Ages, perfect for testing your weaponry.

This is a book for knights-in-training both young and old. It pushes the laws of physics, inspires creativity, proposes experimentation, and fuels the imagination. Most of these siege weapons are great representations of their real-life counterparts, but they cost only pennies, making them perfect for group exercises and outfitting an army.

Keep in mind that this book is for entertainment purposes only. ***Please review the "Play It Safe" page for your personal protection.*** Build and use these projects at your own risk.

PLAY IT SAFE

Those who are granted weapons of great power are commanded to wield them justly. When building and firing your MiniWeapons, be responsible and take every safety precaution. Switching materials, substituting ammunition, assembling improperly, mishandling, targeting inaccurately, and misfiring could all cause harm. Like a guard at the gate, you should always be prepared for the unknown. **Eye protection is a must** if you choose to experiment with any of these projects. In addition, some projects in this book use plastic tableware–knives, spoons, and forks. **Never substitute metal tableware!**

Always be aware of your environment, including spectators and flammable materials, and be careful when handling the launchers. Arrows have sharp points and elastic shooters fire projectiles with unbelievable force. Ammo, no matter what the material, can cause harm. **Never point a launcher at people, animals, or anything of value.** And **never** take or transport any of these projects on public transportation, such as an airplane, bus, or train–these projects are to be used at home.

Remember that because miniweaponry is homebuilt, it is not always accurate. Basic target blueprints and proposed printouts are available at the end of the book and at www.JohnAustinBooks.com. Use these–*not* random targets–to test the accuracy of your MiniWeapon.

Some of the projects outlined in this book require tools such as hobby knives, pocketknives, hot glue guns, wire cutters, and electric drills that can cause injury if handled carelessly. Tools need your full attention—make safety your number-one priority. If you have trouble cutting, your knife may be dull or the selected material may be too hard; stop immediately and substitute one of the two. ***Junior knights-in-training should always be assisted by an adult when handling potentially harmful tools.***

Always be responsible when constructing and using miniweaponry. It is important that you understand that the author, the publisher, and the bookseller cannot and will not guarantee your safety. When you try the projects described here, you do so at your own risk. They are *not* toys!

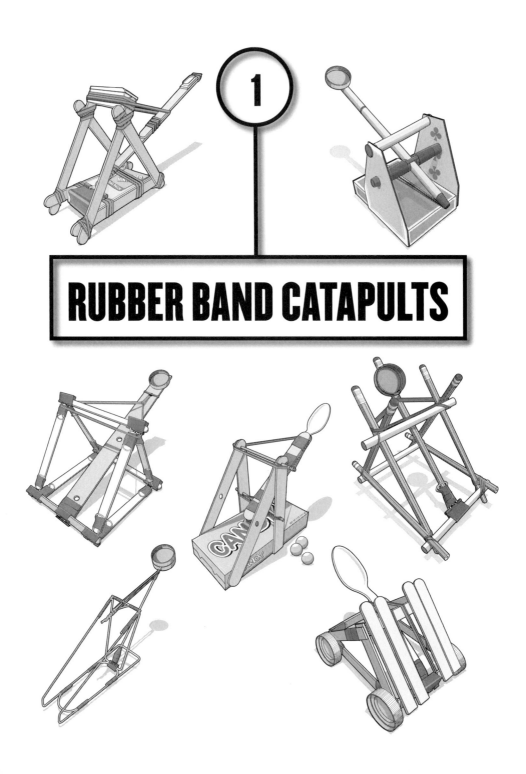

RUBBER BAND CATAPULTS

TIC TAC CATAPULT

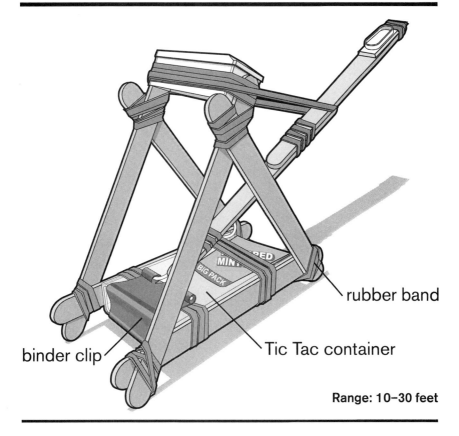

rubber band

binder clip

Tic Tac container

Range: 10–30 feet

As war drums sound in the distance and resources run scarce, this ingenious catapult design could be the kingdom's only salvation. It's the Tic Tac Catapult, and it's awesome! Despite its small size, it has an impressive range and has proven to be very effective during warfare. Plus, it can be quickly constructed without the aid of tools. All the materials needed for assembly can be purchased in bulk, making it easy to mass produce these little siege engines—and doing so won't cost you the royal treasury!

Supplies

1 medium binder clip (32 mm)
1 Tic Tac container
13 rubber bands
8 craft sticks
Duct tape (optional)

Tools

Safety glasses

Ammo

1+ soft candies or small, hard candies

Step 1

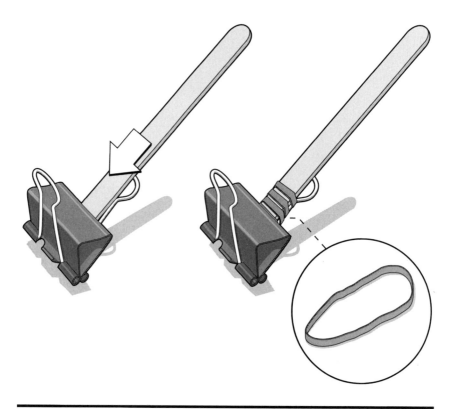

Begin construction with the catapult arm or beam. Locate one medium (32 mm) or similar sized binder clip that can clamp onto a Tic Tic container—test fit it now. Use one rubber band to fasten one craft stick onto the inside of the metal handle attached to the binder clip. Duct tape can be substituted or added for additional strength.

Step 2

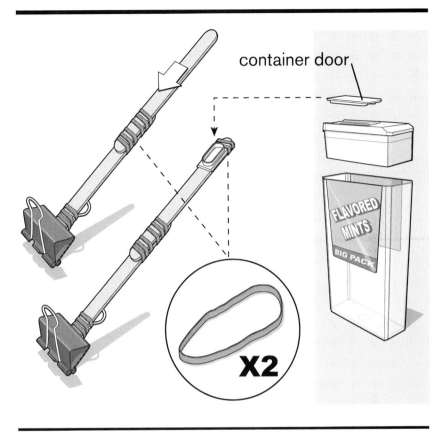

container door

X2

To lengthen the arm, add an additional craft stick to the far end of the fixed stick using one rubber band to secure it into place.

At the end of a catapult arm is the projectile basket, which is fashioned out of a Tic Tac container door. Carefully snap the small door off the lid as illustrated above. Place the door smooth side down at the end of the second craft stick and secure it with a rubber band. The throwing arm is now complete.

Step 3

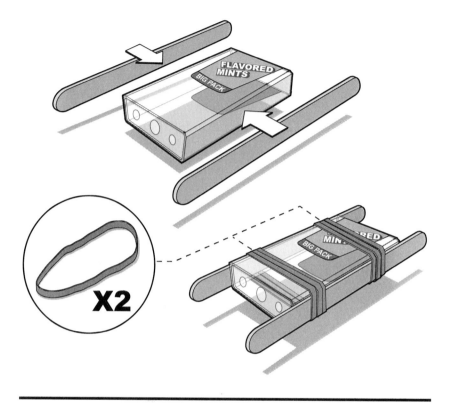

It's time to build the frame out of the Tic Tacs' main container. Fasten two craft sticks to the opposite sides of the container using two rubber bands. The elastic pressure from the two rubber bands may cause some warping at the open end of the container, but minimal warping won't affect the catapult's performance.

Step 4

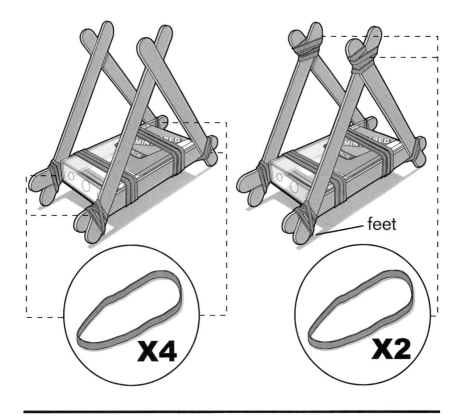

feet

X4 X2

This step requires four craft sticks and six rubber bands. With the foundation of the frame complete, fasten each of the four craft sticks to the endpoints of the container assembly as shown. Once fastened, rotate the sticks to form a triangular frame, then fasten each pair together with one rubber band each. The round ends of these crafts sticks should extend slightly at the bottom to create four feet.

Step 5

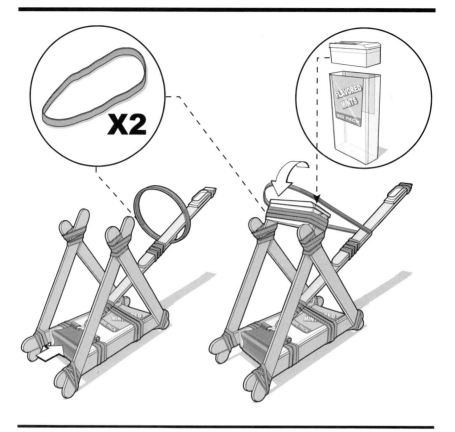

It's time to combine both assemblies. Clamp the arm assembly onto the end of the Tic Tac container frame. Once fastened, the swing arm assembly should move back and forth between the two triangular wooden supports.

This catapult will be powered with one rubber band. Place the rubber band around the swing arm and then slide one end between the two attached craft sticks.

Finally, place the remaining part of the Tic Tac container lid between the two triangular frames to create what is called the padded beam or bed. Fasten the lid to the frame using one rubber band. Then wrap the rubber band fixed to the arm around the cap.

The catapult is complete! When firing, place your hand on the frame for support. *Remember to use eye protection! Never aim this catapult at a living target and use only safe ammunition.* Soft mints, hard candies, and mini marshmallows work nicely.

CARD BOX CATAPULT

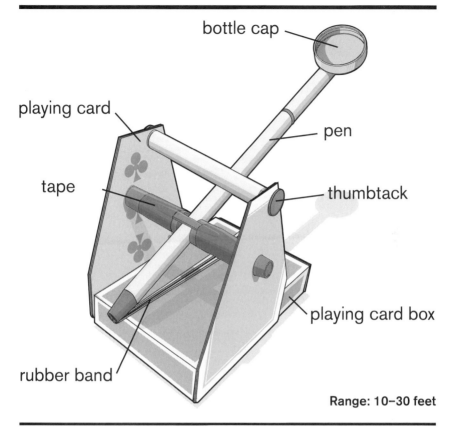

bottle cap

playing card

tape

pen

thumbtack

playing card box

rubber band

Range: 10–30 feet

Game night just got interesting—this MiniWeapon is engineered from laminated playing cards and ballpoint pens. When it's your turn, break out the Card Box Catapult for the ultimate game changer!

Supplies

2 plastic ballpoint pens with caps
1 playing card box
7 playing cards
3 thumbtacks
1 rubber band
1 plastic bottle cap
Duct tape

Tools

Safety glasses

Pliers or thin dowel (optional)
Large scissors or hobby knife
Hot glue gun
Pencil
Single-hole punch
Paper clip (optional)

Ammo

1+ soft candies or similar
objects

Step 1

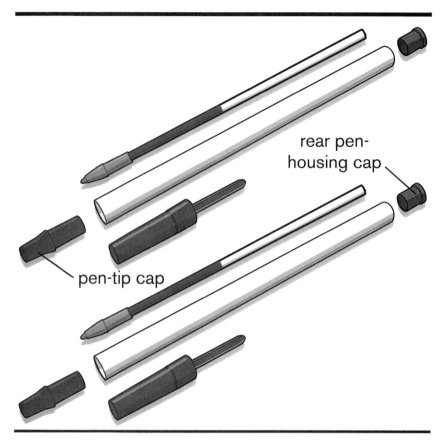

rear pen-housing cap

pen-tip cap

Disassemble two plastic ballpoint pens into their various parts. Dislodge the rear pen-housing cap on both pens. You may need a tool—pliers or a thin dowel—to dislodge the rear pen-housing caps. Also remove the pen-tip caps. Lay out all the components of the pen and do not discard anything at this point.

Step 2

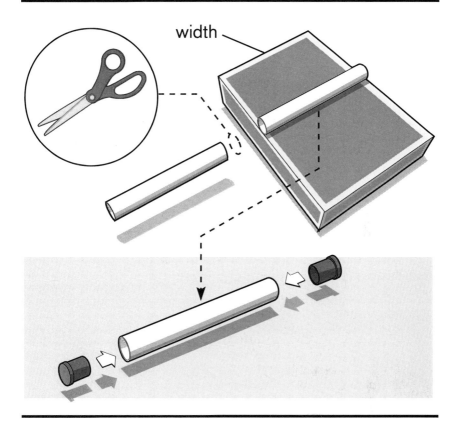

width

Using large scissors or a hobby knife, shorten one of the pen hous-ing tubes to the width of the playing card box; use the box as a guide before cutting.

Take the newly shortened pen housing—the width of the card box—and attach one rear pen-housing cap to each end. Push the caps in completely. The pen housing should now be sealed from both ends.

Step 3

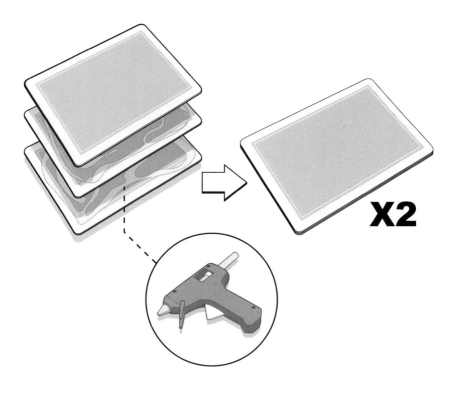

Use a hot glue gun to attach two sets of three playing cards together. These combined sets will make up the two side supports of the catapult. Let both glued card sets dry completely before proceeding to the next step.

Step 4

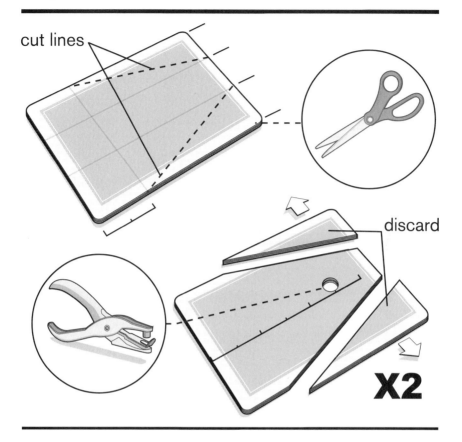

cut lines

discard

X2

Both card stacks will now be transformed into the sides of the catapult. This wedge design is only for aesthetics, so exact measurements are not essential. The following measurements are provided only as a guide.

To create the wedge, establish the bottom of the card supports and pencil a mark roughly ½ inch up on both sides. On the opposite end of the cards, divide the width into three equal sections, and mark each third with a pencil. These marks will be the guide for cutting. With the scissors or hobby knife, use these marks to remove both wedges from the card assembly.

Next, measure 2½ inches up from the uncut end of the card, then punch one center hole using a single-hole punch. A hobby knife can be substituted if needed.

Repeat both steps on the second card stack. You should have two identical card stacks with aligned holes when completed.

Step 5

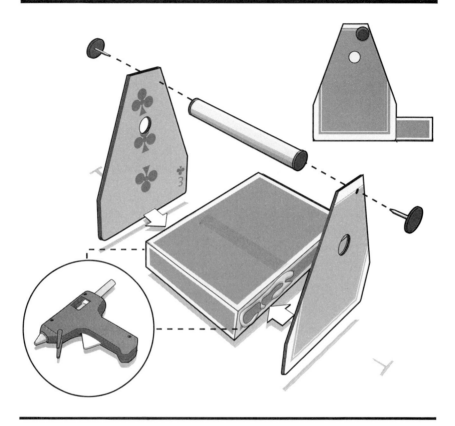

Apply hot glue to both sides of the card box and place both card stacks flush with the bottom of the box. It is important that the punched holes line up.

When the glue is dry, place the modified pen assembly between the two card stack supports and hold it in place using thumbtacks at both ends. Use the side-view drawing above for reference. Once in place, the attached pen assembly serves as the catapult's padded beam that will catch the swing arm.

Step 6

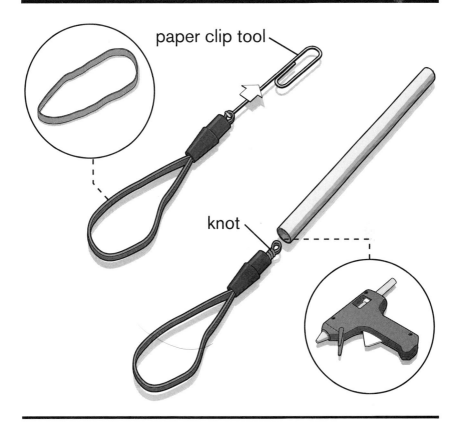

paper clip tool

knot

Feed one rubber band through the pen tip. You may need a tool help feed the band; to make one, bend a small hook at the end of a paper clip to assist with this delicate task. On the larger-diameter side of the pen tip, tie a knot into the rubber band. This knot should prevent the rubber band from passing back through the pen tip. Tie additional knots if needed.

Next, hot glue the pen tip assembly into the uncut pen housing. The glue will prevent the pen tip from dislodging while firing the catapult and reinforce the fixed rubber band.

Step 7

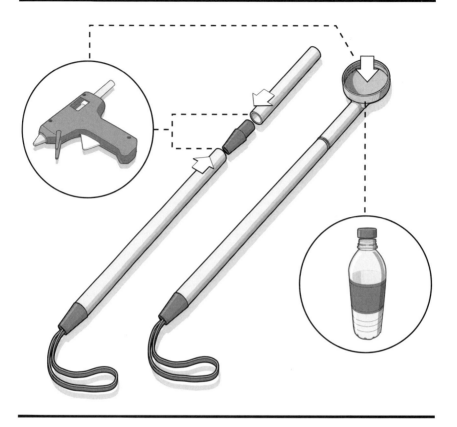

On the opposite end of the pen housing assembly, hot glue and plug the remaining pen tip into the housing as shown. When the glue dries, hot glue and plug the remaining pen housing you cut and removed in step 2. These three parts should be aligned when you're finished.

Hot glue one plastic bottle cap from a soft drink bottle to the swing arm assembly, opposite the rubber band. Let the glue dry before handling.

Step 8

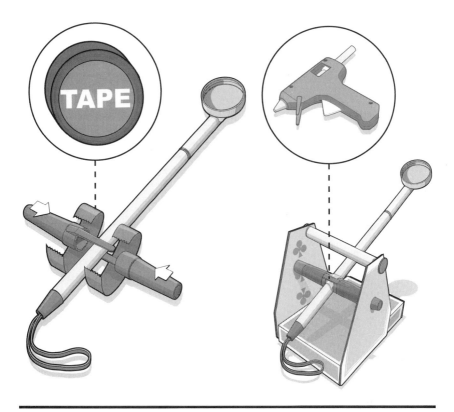

It's time to complete the swing arm assembly by adding the rotation beam crafted out of two pen-tip caps. This step will require two hands and a little patience, so it's best if you precut two pieces of duct tape roughly 1½ inches long before starting.

Perpendicular from one another, and no more than 2 inches from the bottom of the pen arm, sandwich two pen caps around the pen housing, as shown above. Once each pen cap's clip overlaps the other cap, tape them together.

Now, slide the completed swing arm assembly between the card frame until both pen cap ends snap into the two punched holes. The swing arm should then rotate back and forth. Adjust the pen cap axis point if needed by sliding the swing arm up and down, and then add additional hot glue to the adjoined pen caps.

Step 9

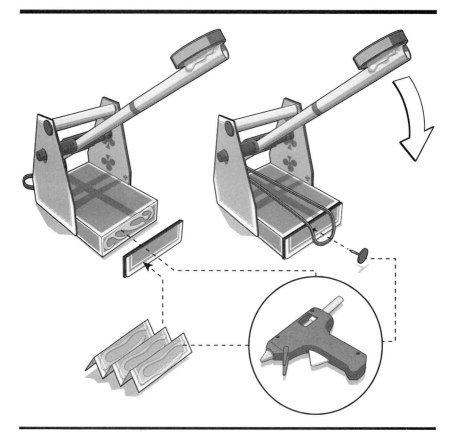

The Card Box Catapult is almost ready for battle. Affix the rubber band on the swing arm to the rear of the card box. To strengthen this connection, fold up one playing card like an accordion and then add hot glue to hold the folds together.

When it is dry, glue the folded-up card to the rear of the card box as shown. Then fix the last thumbtack to the center of the folded up card, with the rubber band looped underneath. For safety reasons, add additional hot glue to strengthen the connection of the thumbtack.

Remember to use eye protection! Never aim this catapult at a living target and use only safe ammunition. Soft mints, candy, and mini marshmallows work nicely.

RULER CATAPULT

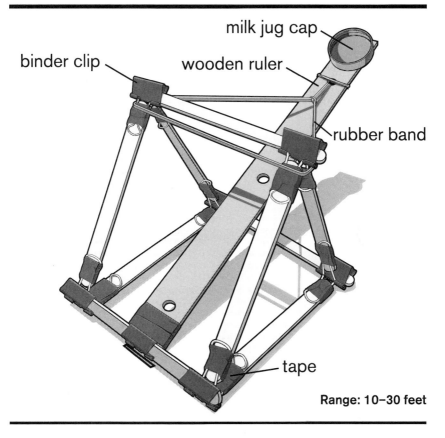

milk jug cap

binder clip

wooden ruler

rubber band

tape

Range: 10–30 feet

When a rival lord locks himself in his fortress and refuses to surrender, a good old-fashioned siege is necessary. The mere sight of the Ruler Catapult, with its formidable foot-long swing arm and metal clasp bracing, will quickly persuade the feuding nobleman to sue for peace. However, if negations go south, there's no need to huff and puff; you can just blast his house down!

Supplies

7 small binder clips (19 mm)
9 craft sticks
Duct tape
1 wooden ruler with peg holes
2 rubber bands
1 plastic milk jug cap

Tools

Safety glasses
Hot glue gun

Ammo

1+ mini marshmallows

Step 1

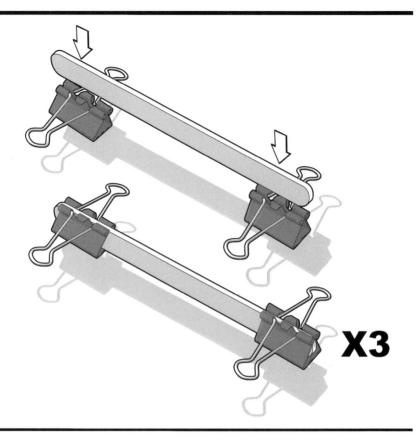

X3

To begin building the Ruler Catapult, you must first assemble three identical braces. Clamp two binder clips at the opposite ends of one craft stick as shown. Repeat this step twice more, so you have three completed braces.

Step 2

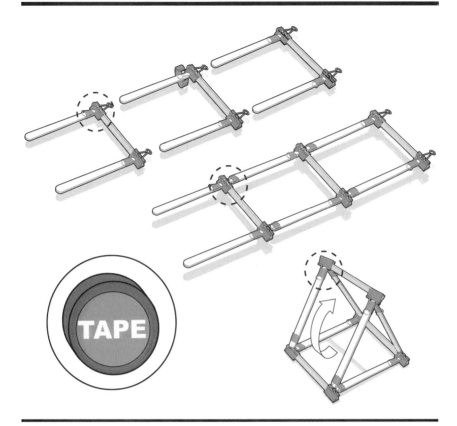

TAPE

Now attach the braces together using six additional craft sticks to build the catapult frame. Add two craft sticks to each brace by taping each stick to a binder clip handle on the same side of the assembly. Repeat this step three times (top image).

Next, with all the binder clip clamp openings facing upward, tape all three of the assemblies together by attaching the binder clip handles (middle image).

Once your frame is complete, fold the assembly into a triangle as shown. Then tape the remaining binder clip handles to the two remaining craft sticks. When completed, this frame should be structurally sound (bottom image).

Step 3

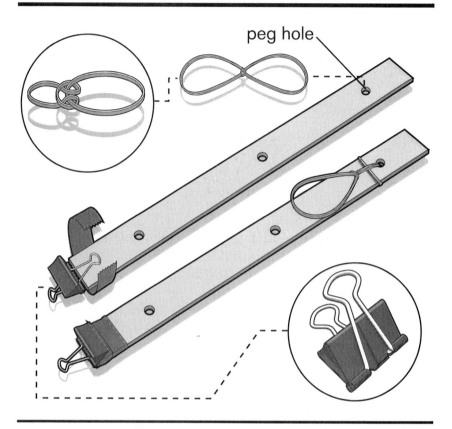

peg hole

The swing arm on this catapult will be constructed from one wooden ruler. However, a plastic or metal ruler could be substituted. Begin by looping two rubber bands together to construct one large double rubber band, as shown. Then slide one end of the rubber band through the ruler's top peg hole and back around the ruler to hold the double rubber band in place.

At the opposite end of the wooden ruler, tape just one of the small binder clip's handles to the ruler, which enables the binder clip to swing back and forth while attached. Do not clamp the binder clip onto the ruler.

Step 4

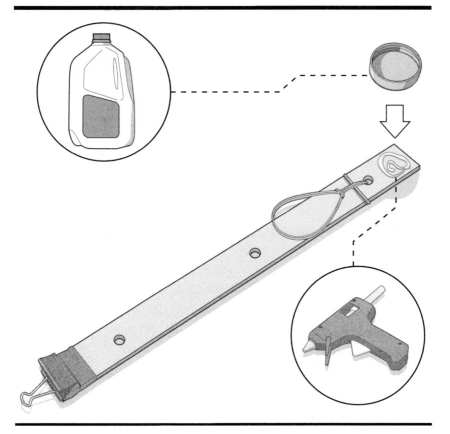

Next, carefully hot glue a plastic milk jug cap (or a cap similar in size) to the end of the ruler, above the double rubber band and opposite the binder clip. You have just completed the catapult arm.

Step 5

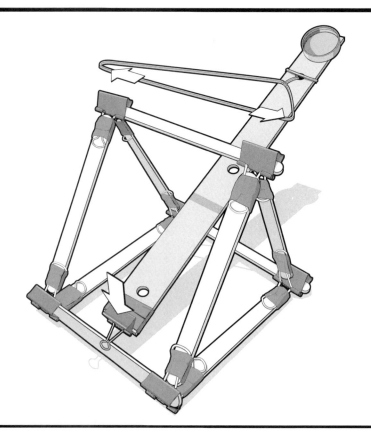

Attach the swing arm assembly to the front of the catapult frame by clamping the attached binder clip to the craft stick front brace. The arm should be centered onto the brace, and swing freely up and down.

To complete the catapult, loop the attached double rubber band around the top of the craft stick frame. To fire, hold the bottom of the frame with one hand, pull back the loaded swing arm, and release when ready!

Remember to use eye protection! Never aim this catapult at a living target and use only safe ammunition. Soft mints and mini marshmallows work nicely.

CANDY BOX CATAPULT

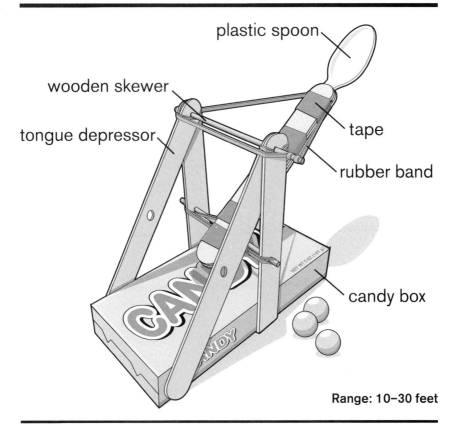

plastic spoon

wooden skewer

tongue depressor

tape

rubber band

candy box

Range: 10–30 feet

Hungry for a victory? The Candy Box Catapult is a sweet choice for your next siege attack! Not only is it engineered to produce tremendous firepower across ample distances on the battlefield, but it also features an all-in-one ammo holder integrated into its base.

Supplies

1 wooden skewer
1 candy box (5 oz. or similar size)
Duct tape
6 wooden tongue depressors
4 rubber bands
1 plastic spoon

Tools

Safety glasses

Wire cutters
Hobby knife
Hot glue gun
Power drill
Large scissors (optional)

Ammo

1+ soft candies

Step 1

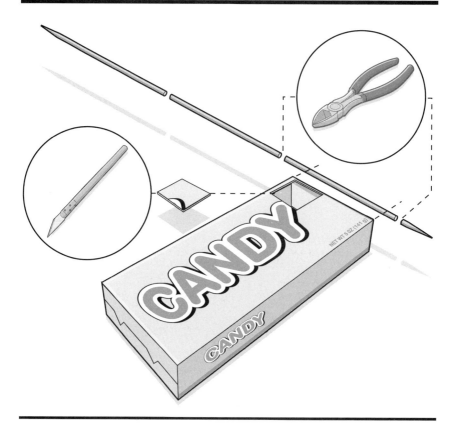

Using wire cutters, carefully cut off two identical sections of a wooden skewer that are both 1 inch longer than the width of the candy box. When cut, the lengths of the wooden skewer sections will overhang the width of the candy box.

Then, with a hobby knife, cut a small door into the top corner of the candy box. If the box has been opened, reseal the open flaps with hot glue or tape for structural reasons.

Step 2

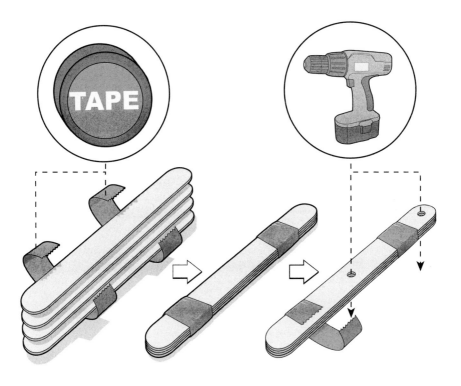

Bundle four wooden tongue depressors using two pieces of tape, as illustrated.

Carefully drill two holes into the stack of depressors using a power drill with a drill bit diameter that is slightly larger than the wooden skewer's diameter. Drill the first hole at the end of the stack. Make sure the hole is an equal distance from the sides and edge of the stick, as shown. Drill the second hold in the center of the stack.

Both holes should be drilled completely through the stack. After the holes have been drilled, remove the tape and separate the tongue depressors.

Step 3

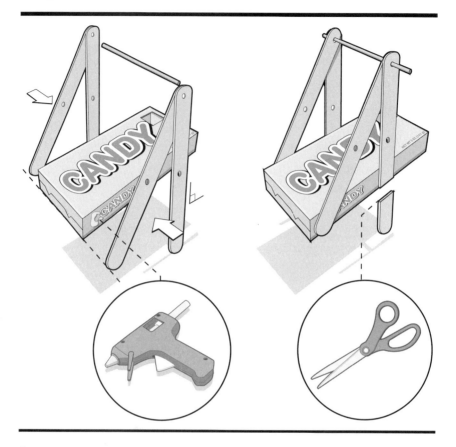

Separate the tongue depressors into two pairs. Then, using the top drilled holes, slide each pair onto the opposite ends of one of the shortened skewer rods.

Once they're in place, hot glue the four wooden depressors to the side of the candy box as shown in the illustration on the right. The front pair of depressors should be glued at a 45-degree angle and flush to the bottom of the box; the second pair of depressors should be glued at 90-degree angle with some material overhanging the box. After the glue is dry, use the hobby knife or a pair of large scissors to clip the excess material off the bottom of the tongue depressors so that the box sits flush with the tabletop.

Step 4

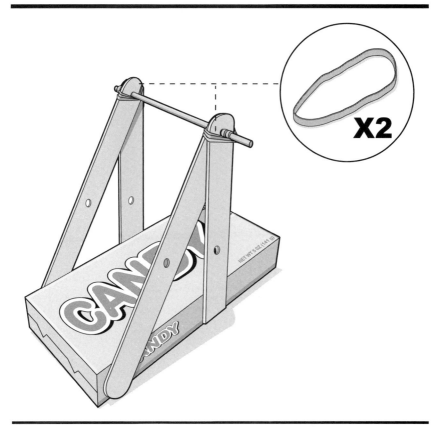

X2

Fasten a rubber band around each of the two outer ends, where the skewer and depressors meet. This will help stabilize the frame and prevent the skewer from dislodging during firing.

Step 5

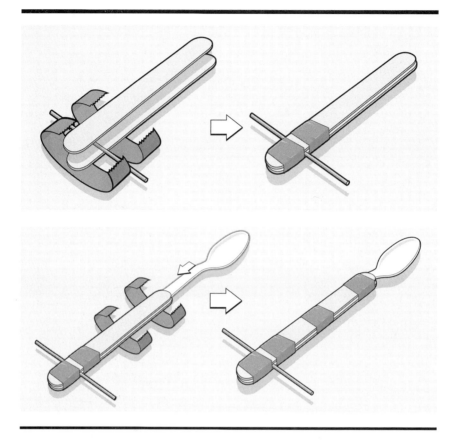

Now it's time to assemble the swing arm. Center the second short-ened skewer in between the far ends of the two remaining tongue depressors. Then sandwich the skewer into place by taping each side around the skewer, as shown in the top illustration.

On the opposite end of the swing arm assembly, slide one plastic spoon between the two tongue depressors and tape it in place by wrapping tape around the tongue depressors (bottom illustration).

Step 6

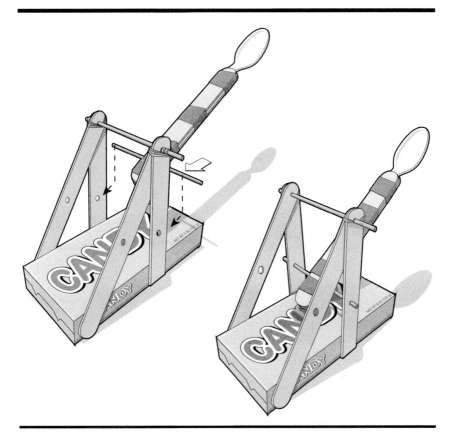

Attach the swing arm assembly to the frame by slightly angling and guiding the attached skewer into the lower center holes on the wooden catapult frame.

The swing arm should swing freely, and the bottom of the arm should clear the top surface of the candy box. If there is resistance between the swing arm and the box, adjust the swing arm skewer.

Step 7

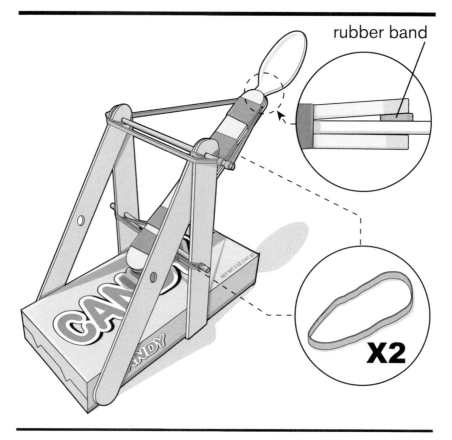

rubber band

X2

It's time to add some elastic firepower to this catapult. To do so, wedge one rubber band between the front tongue depressor and the spoon neck. Then loop the other end of that rubber band around the top of the frame. Depending on the size of the rubber band and desired tension, you might need to wrap the rubber band around the skewer more than once.

Place the last rubber band around the lower skewer. This will prevent the skewer from sliding out of the frame and provide additional support.

Now you're ready to fire. **Remember to use eye protection! Never aim this catapult at a living target and use only safe ammunition.** Soft mints and mini marshmallows work nicely, and the door you cut into the candy box in step 1 allows you to store the extras in the catapult's base.

PEN AND PENCIL CATAPULT

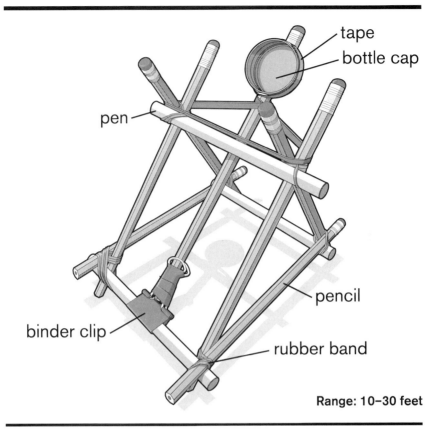

tape

bottle cap

pen

binder clip

pencil

rubber band

Range: 10–30 feet

The Pen and Pencil Catapult has all the classic elements of a formi-dable desktop catapult. Fabricated from basic office supplies, it's the perfect siege weapon for a rainy day. Display it with pride and those who oppose you will suffer grave consequences.

Supplies

7 wooden pencils
13 rubber bands
3 plastic ballpoint pens
1 plastic bottle cap
1 large paper clip (length must be greater than the diameter of the bottle cap)

Duct tape
1 small or medium binder clip (19 mm or 32 mm)

Tools

Safety glasses
Pliers

Ammo

1+ soft candies

Step 1

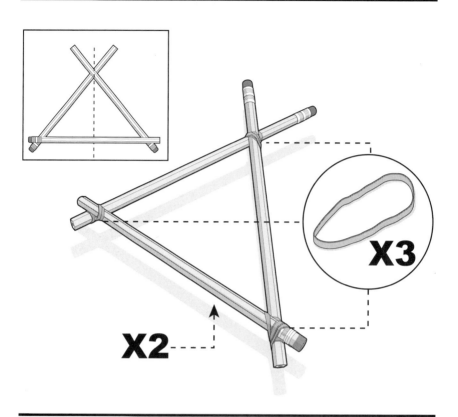

Construct a triangle frame out of three pencils, securing the connections with three rubber bands. The finished triangle will have three roughly equal sides. The pencils should overhang at the ends; these sections will become the catapult's feet.

Repeat this step with another three pencils so you have two identical triangles.

Step 2

X6

Combine both triangle assemblies from step 1 into a single frame by adding three plastic ballpoint pens as cross braces. First, remove the ink cartridges from the pens (optional). Then attach each pen to one of the corners of the pencil assembly using one rubber band on each end to secure it in place.

Continue fastening the pens to both triangles while keeping the triangles aligned with one another. You might have to bend the finished frame to straighten out the design.

Step 3

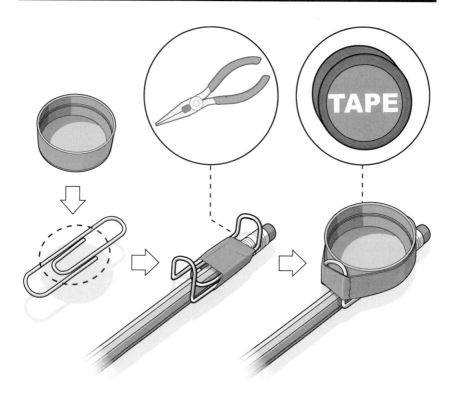

With pliers, make two 90-degree bends in the large paper clip, using the bottle cap as a guide: the distance between the bends should be the same as the cap's diameter. The bent paper clip will act as a cradle for the bottle cap and will add much-needed support to the swing arm basket.

Now tape the modified paper clip to the end of a pencil. Once in place, sandwich the plastic cap between the paper clip bends and tape the cap into place.

Step 4

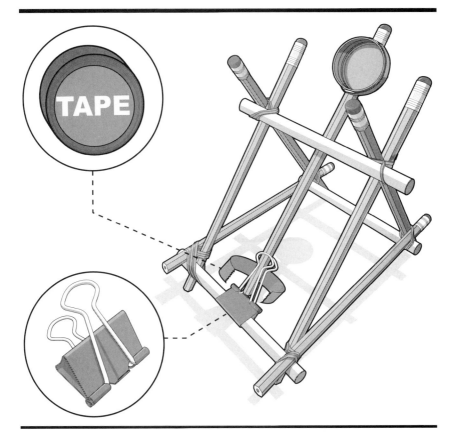

Complete the swing arm assembly by clamping a small or medium binder clip to the center of the lower pen fixed to the triangular catapult frame.

Once in place, tape the swing arm assembly from step 3 to the metal binder clip handles with the bottle cap facing forward. The swing arm should rotate freely between the two pens.

Step 5

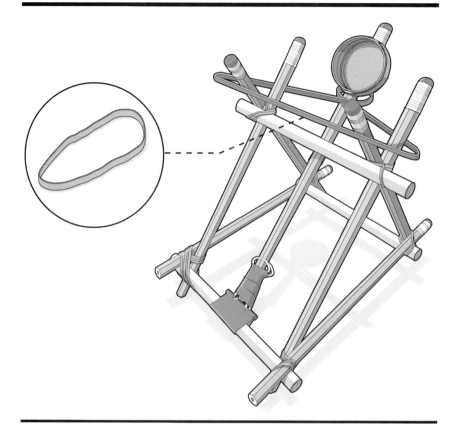

Now loop the last rubber band around the neck of the swing arm assembly, right below the catapult basket (bottle cap). Loop the other end of the rubber band around the upper frame of the catapult as pictured. This desktop catapult is complete!

Remember to use eye protection! Never aim this catapult at a living target and use only safe ammunition. Soft mints and mini marshmallows work nicely.

METAL HANGER CATAPULT

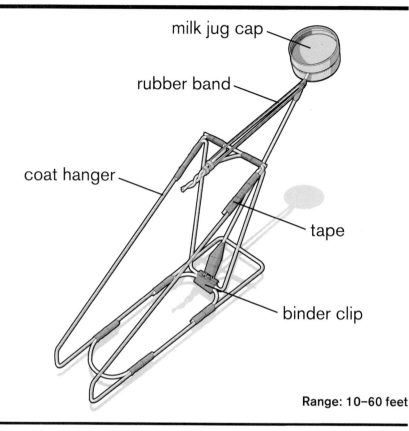

milk jug cap

rubber band

coat hanger

tape

binder clip

Range: 10–60 feet

Fabricated from bent metal coat hangers, the Metal Hanger Catapult can't be burned by the flaming weapons you'll frequently face during a siege. Its nontraditional design is sleek and sturdy, with a straightforward firing mechanism perfect for bombarding enemy structures and territory. After a few quick steps, you can rock and roll this metal machine to victory.

Supplies

2 metal coat hangers
Duct tape
1 small binder clip (19 mm)
2 plastic milk jug caps (or similar)
1 rubber band

Tools

Safety glasses
Pliers
Wire cutters
Hot glue gun

Ammo

1+ soft candies

Step 1

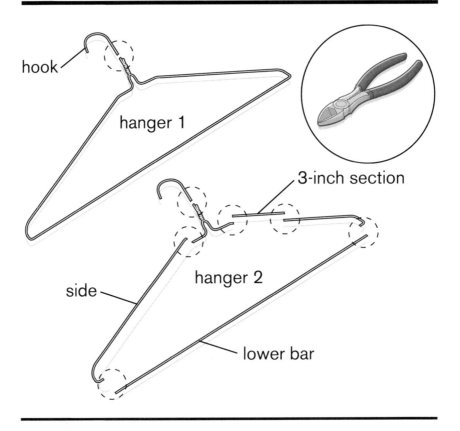

hook

hanger 1

3-inch section

hanger 2

side

lower bar

With a pair of pliers or wire cutters, remove the hook end from two metal hangers as illustrated. Hanger 1 is now complete.

From hanger 2, remove the lower bar by cutting it at both ends. Next, remove one side of the hanger by cutting below the twisted neck detail. On the opposite side, remove a 3-inch-long section of the coat hanger. Use the illustration above for reference.

Step 2

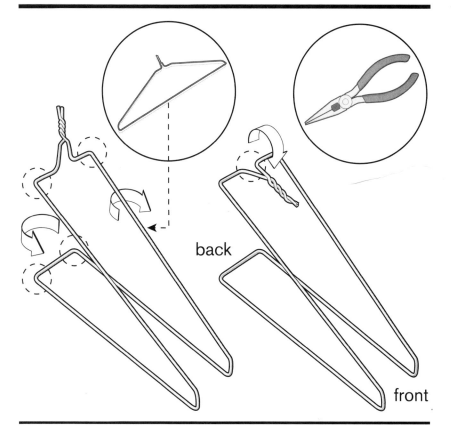

back

front

With pliers, fold hanger 1 in four spots so that when finished, the modified hanger sits upright and is approximately 2 inches wide—about the same width as the removed hook.

Next, bend the neck forward so that the spiral section is level with the angled frame.

Step 3

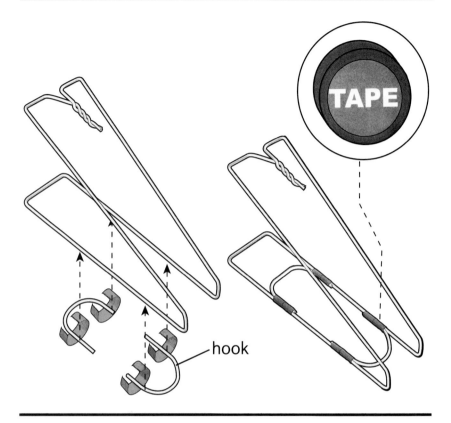

hook

Tape both of the removed hooks to the bottom of the metal frame. The first hook should be placed toward the front of the frame assembly on the base; its location can be approximate. However, the second hook should be placed about 1 inch forward of the top rear edge of the frame assembly. See the illustration for placement.

Note: Once this catapult is completed, the second hook can be adjusted to match the shooter's preference for height and range.

Step 4

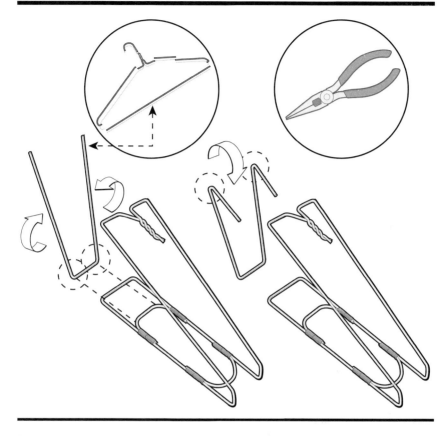

To support the rear of the wire frame, bend the removed straight bar from hanger 2 into a "U" shape as shown, with the bottom of the "U" bend approximately the same width as the frame assembly.

The placement of the next two bends is determined by the height of the frame assembly. Use the frame as a guide. Then, with pliers, bend the two ends of the "U" section more than 90 degrees toward the front. This angle should align with the frame assembly's angle.

You will mount this piece in the next step, so adjustments can be made then if necessary.

Step 5

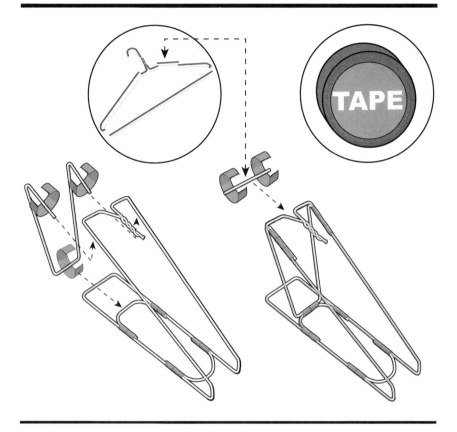

Tape the lower bar of the modified "U" to the back of the frame assembly. Place tape at all three spots where the bars join, two pieces for the top and one piece for the bottom.

Next, tape the 3-inch rod to the top rear of the frame assembly. This rod will be the new swing arm stop and will improve the accuracy of the swing arm by blocking the preexisting gap in the frame.

Step 6

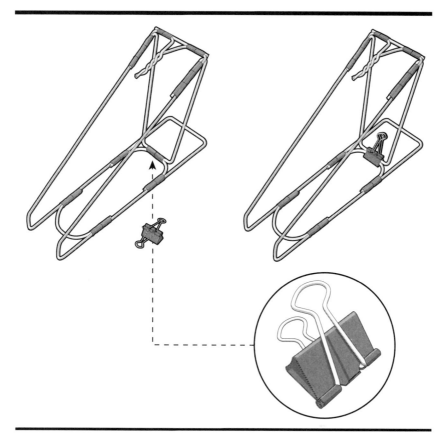

Now snap a small binder clip to the bottom center brace of the frame assembly, where the rear hook and modified "U" meet. After the binder is clamped on and centered, extend its two metal handles upward.

Step 7

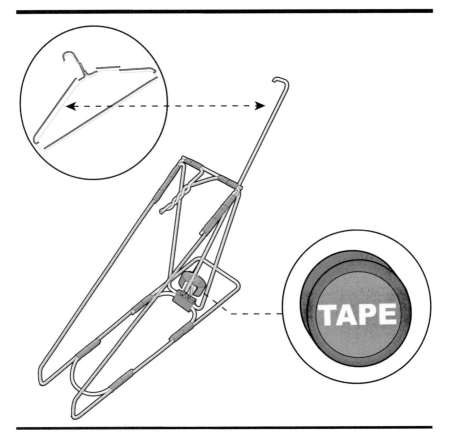

The swing arm will be made out of the remaining side section that was cut from hanger 2. Tape the straight end of this section between the two metal handles of the attached binder clip. Add plenty of tape to prevent the swing arm from coming loose during firing.

Step 8

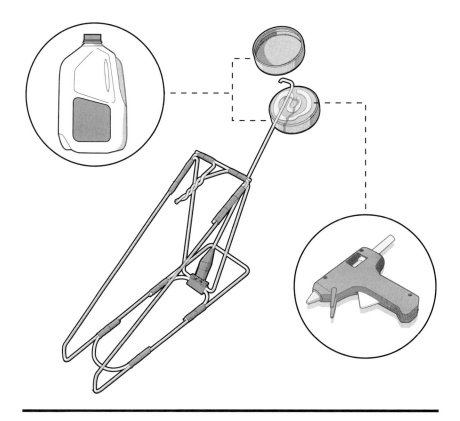

The catapult basket–where the ammo is loaded and launched–will be made from two plastic milk jug caps.

Using hot glue, carefully sandwich both caps, smooth sides together, around the metal tip of the swing arm. Why two caps? The pair of glued caps will prevent the basket from popping off and launching with the ammo, which might happen if you use only one plastic milk jug cap.

Step 9

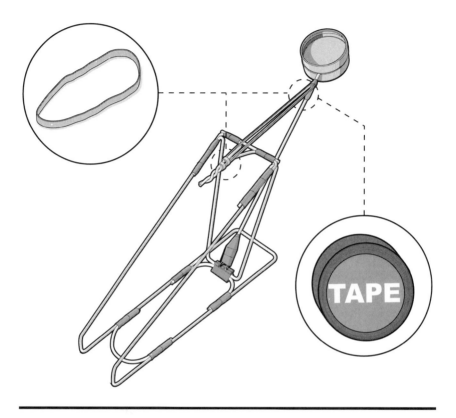

TAPE

Loop the rubber band around the neck of the swing arm, right below the plastic cap. Tape the rubber band in place to prevent it from sliding down the swing arm.

Loop the other end of the rubber band around the neck detail bent in step 2. If necessary, add tape to prevent it from slipping off.

Before heading off to battle, remember the importance of safety. ***Eye protection is a must! Never aim this catapult at a living target and use only safe ammunition.*** Soft mints and mini marshmallows work nicely.

ARMORED CATAPULT

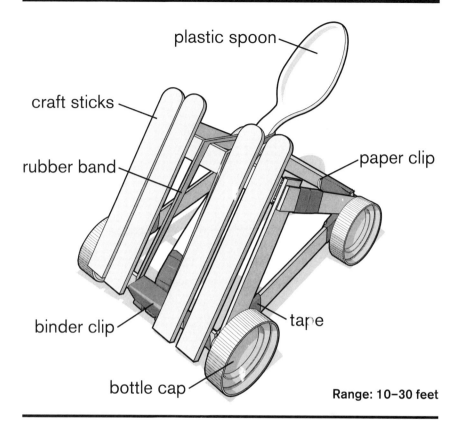

plastic spoon

craft sticks

rubber band

paper clip

binder clip

tape

bottle cap

Range: 10–30 feet

Wake up the bagpipers! It's time to expand the kingdom! During the chaotic offensive, the Armored Catapult will provide critical protection for longbowmen and other missile-launching forces. Decked out in wood-plated armor, it's designed to get you safely within range of the castle so you can unleash all types of wall-crumbling projectiles.

Supplies

6 large paper clips
12 craft sticks
Duct tape
1 small binder clip (19 mm)
1 plastic spoon
1 rubber band
4 plastic bottle caps

Tools

Safety glasses
Pliers
Large scissors
Hot glue gun

Ammo

1+ soft candies

Step 1

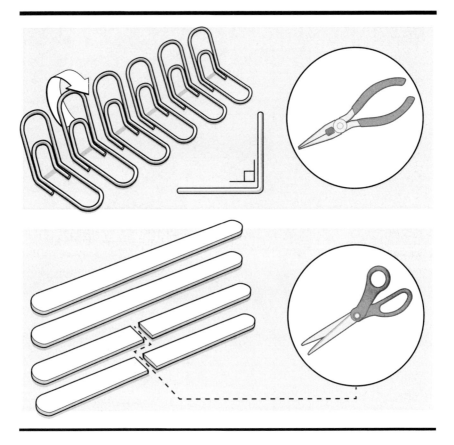

With pliers, bend a 90-degree angle at the center of six large paper clips. Four of these clips will be used in step 2; save the two remaining clips for step 4.

Next, prepare four craft sticks for the base assembly. Carefully cut two of these craft sticks in half with large scissors. Save one of these cut craft sticks for step 4.

Step 2

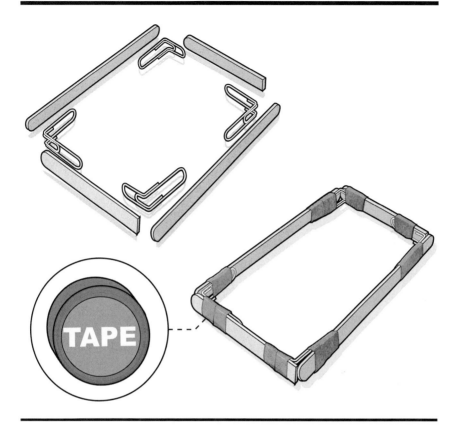

TAPE

Build a box frame using two full craft sticks and two of the cut halves. Join the sticks together by placing four 90-degree paper clips in the interior corners of the box and then fastening this assembly using eight pieces of tape, as pictured.

Step 3

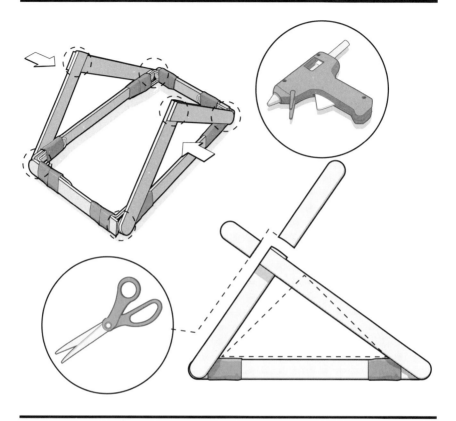

With the box frame completed, construct the vertical frame using four uncut craft sticks. Starting on one side, carefully hot glue two craft sticks together onto the frame to create a triangle as shown—with no sides equal. Repeat this step on the opposite side, aligning the triangles.

When the vertical frame is complete, use the large scissors to trim the excess from the top of both triangles, as shown.

Step 4

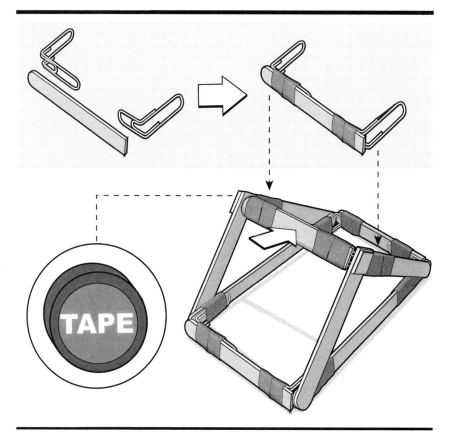

Similar to the frame construction in step 2, fasten two 90-degree paper clips to the ends of a craft stick half-section (cut in step 1) to make a brace.

Slide this brace between the top of the triangle frames and secure it with tape. The frame is now complete.

Step 5

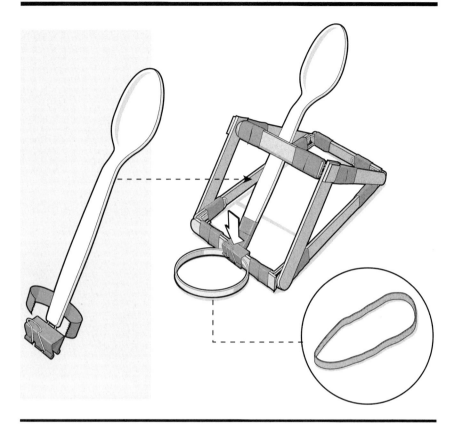

With a small binder clip and a plastic spoon, you will construct the plastic swing arm.

To start, tape one of the metal handles of the binder clip to the underside of the plastic spoon, as illustrated at the left. (Do not remove the second metal handle of this clip.) The swing arm is complete.

Next, clamp the swing arm assembly to the lower craft stick brace on the frame. Once it is centered on the brace, slide one rubber band under the binder clip, as shown. Then remove the front metal handle of the binder clip.

Step 6

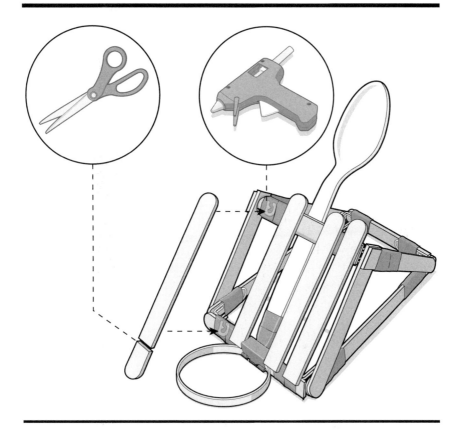

It's time to add some protective armor! Use the large scissors to trim off the round ends from the bottom of four wooden craft sticks, then carefully hot glue all four sticks to the front two braces of the catapult frame as shown.

Step 7

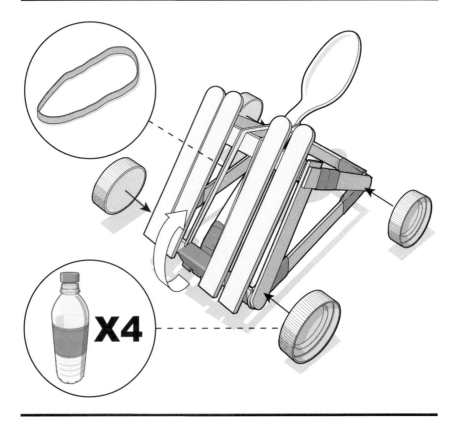

Add some realism to this desktop replica by carefully hot gluing four plastic bottle caps to the corners of the catapult frame to simulate wheels. This is also a quick and easy way to level the catapult. Additional customization is encouraged if you want your catapult model to roll. You might even use brown chocolate milk jug caps for the authentic wooden-wheel look.

Remember to use eye protection! Never aim this catapult at a living target and use only safe ammunition. Soft mints and mini marshmallows work nicely.

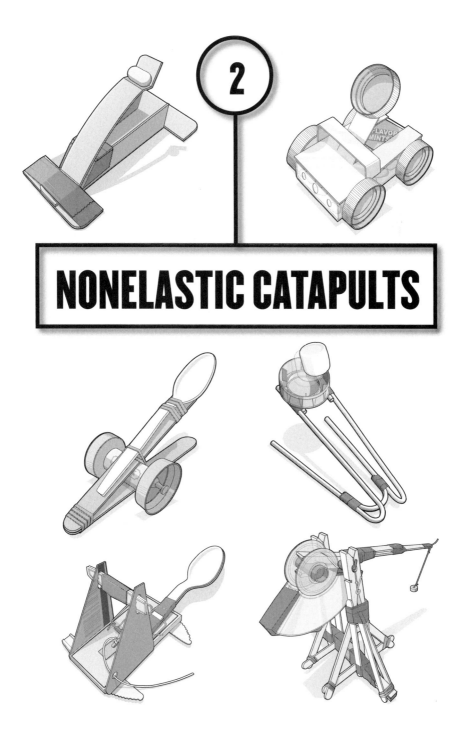

NONELASTIC CATAPULTS

GIFT CARD CATAPULT

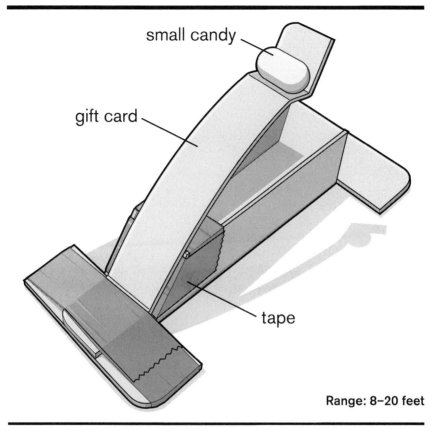

small candy

gift card

tape

Range: 8–20 feet

Manufactured from a depleted gift card, this MiniWeapon is perfect for being indiscreet. Tribute-strapped warlords will drool at the Gift Card Catapult's simplicity while they witness its projectile-hurling potential. Plus, assembling this siege weapon requires only tape and scissors.

Supplies

1 expired or zero-balance plastic gift card
Duct tape

Tools

Safety glasses
Scissors
Permanent marker

Ammo

1+ soft candies

Step 1

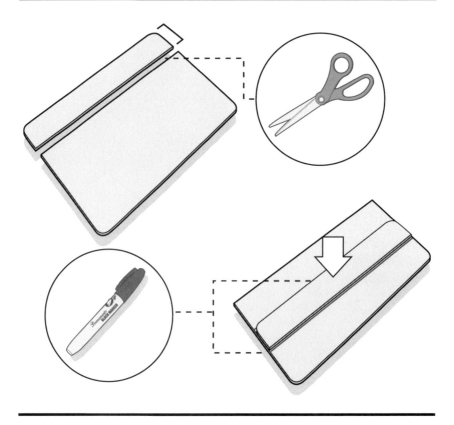

With scissors, cut one ½-inch strip off the long side of an old gift card.

Next, center the ½-inch strip on top of the leftover gift card piece, as illustrated on the right. Use a marker to trace guidelines down both edges of the centered smaller strip.

Step 2

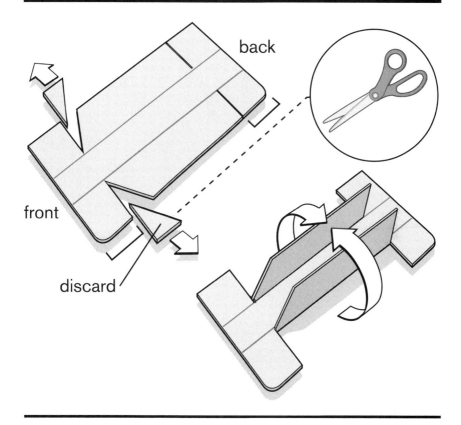

back

front

discard

To construct the frame, make several small cuts into the larger card. First, cut four small lines up to the marker guidelines. Each of these cuts should be ½ inch from the front and back ends of the card. Then, on the front, cut a small ¼-inch wedge toward the back of the card and then discard the removed plastic wedges.

Next, using the guidelines, fold the center of the card upward to a 90-degree angle, forming two walls parallel with one another.

Step 3

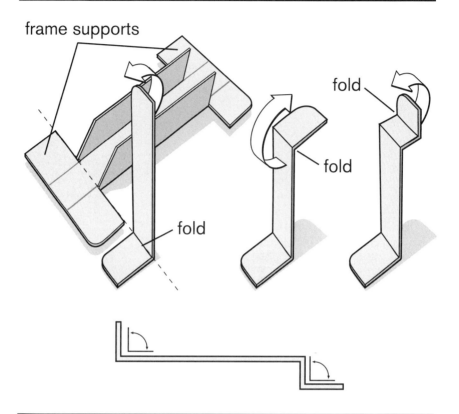

frame supports

fold

fold

fold

fold

The catapult swing arm will be constructed from the ½-inch strip you removed in step 1. The strip will be folded three times in this step in order for the swing arm to function properly.

Make the first 90-degree fold the width of the frame supports. Create the next 90-degree fold on the opposite end of the strip, ¾ inch from the end. Finally, make the last 90-degree fold in the center of that ¾-inch section of swing arm and parallel to the main swing arm.

Use the illustration on the bottom as a guide. Notice how the finished folds are 90-degree steps.

Step 4

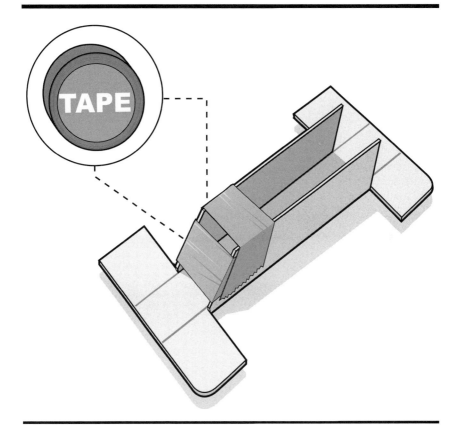

Place two small pieces of tape over the front angled supports and front top surface of the frame. This tape will provide a stop for the swing arm, allowing the plastic swing arm added in the next step to bend backward and then spring forward to launch projectiles.

Step 5

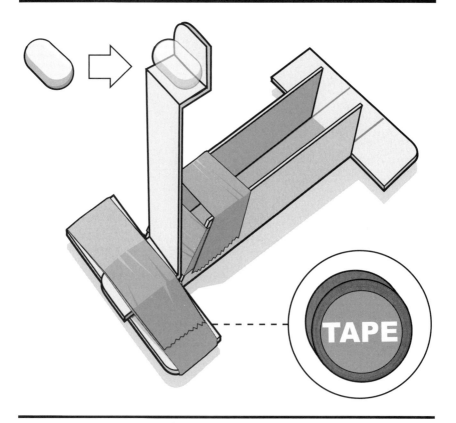

TAPE

It's time to assemble the catapult. Place the swing arm's lone 90-degree bend on top of the front frame support, in front of the taped angle, as shown. Then secure the swing arm by wrapping tape around the front support and over the swing arm.

Test your Gift Card Catapult by placing a small candy on the swing arm step detail. This makeshift basket will hold the ammo in place as you pull down and release the swing arm.

Remember to use eye protection! Never aim this catapult at a living target and use only safe ammunition. Soft mints, pencil erasers, and mini marshmallows work nicely.

TIC TAC ONAGER

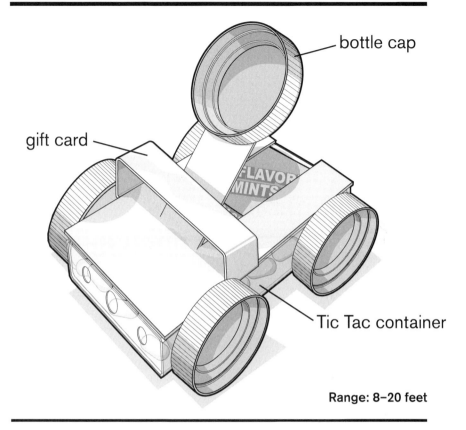

bottle cap

gift card

Tic Tac container

Range: 8–20 feet

With a design borrowed from the legions of the Roman Empire, the Tic Tac Onager is a mighty little catapult that gains its power from a flexible swing arm, not twisted rope like its big brother, making construction a whole lot easier. The solid frame construction will outlast any rigorous campaign to expand your desktop domain.

Supplies

1 Tic Tac container

1 expired or zero-balance plastic gift card

5 plastic bottle caps

Tools

Safety glasses

Scissors

Hot glue gun

Ammo

1+ small, hard candies or mini marshmallows

Step 1

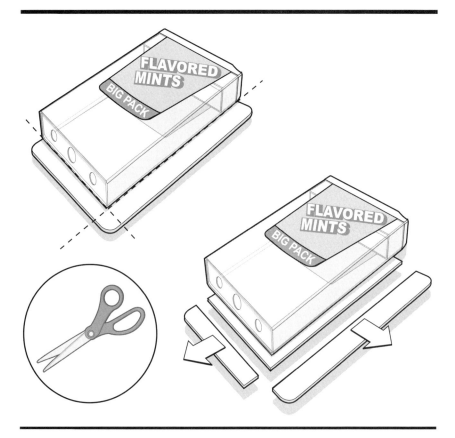

Place a Tic Tac container on top of an expired or zero-balance plastic gift card. Square it to one corner and cut any card material that overhangs with scissors.

Do not discard the scraps; the long card strip will be used to make the catapult's padded beam in step 3.

Step 2

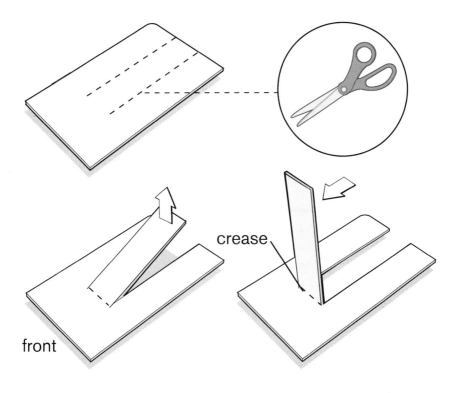

crease

front

With scissors, cut two slits into the center of the card approximately ½ inch from each side. Each slit should end ¾ inch from the front of the card.

Now bend the section of the card between the slits to a 90-degree angle and crease the plastic. This flap will be the spring mechanism for the catapult.

Step 3

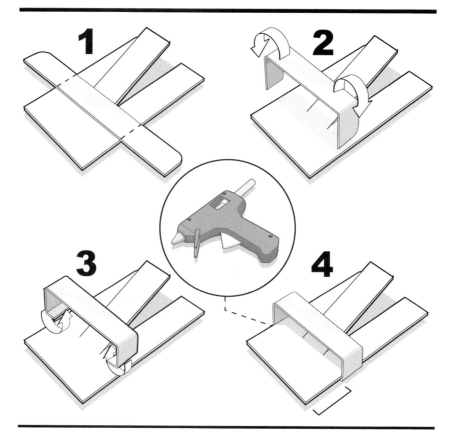

Center the long card strip on top of the gift card as shown (1). Next, fold two 90-degree angles into the strip where it overhangs the card. The folded strip should now be the same width as the bottom card (2).

Now fold two more 90-degree angles, roughly ¼ inch from the ends of the folded strip (3). Place the folded strip ¾ inch from the front of the card, aligned and overhanging the cut slits, and then carefully hot glue the small folded tabs to the underside of the card (4). This added detail will stop the swing arm and assist in the launching of projectiles.

Step 4

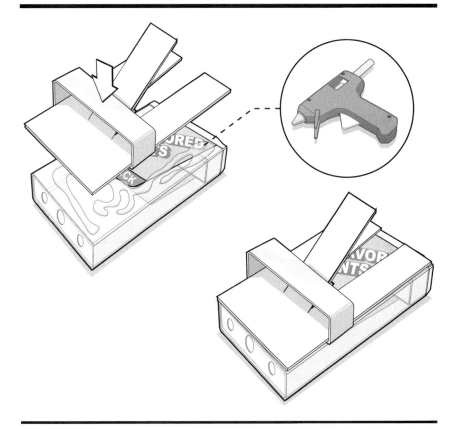

Position the gift card assembly on top of the Tic Tac container. The swing arm should be on the same side as the container's door. Carefully hot glue the assembly to the container, but *do not* glue the swing arm.

Step 5

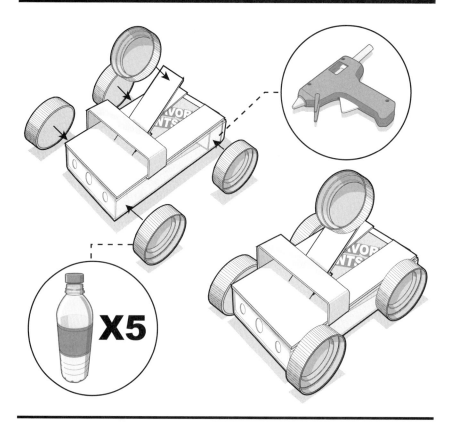

For the last step, attach four wheels and a catapult basket. Carefully hot glue four plastic bottle caps to the corners of the Tic Tac container. Although these wheels do not roll, they will add authenticity to the catapult. (Additional customization is encouraged if you want the catapult model to roll.)

Now glue the remaining bottle cap to the top of the swing arm. If a fifth cap is not available, the Tic Tac container door can be removed and glued on the swing arm for similar results.

When firing, ***remember to use eye protection! Never aim this catapult at a living target and use only safe ammunition.*** Hard candies and mini marshmallows work nicely.

DEPRESSOR SPOON CATAPULT

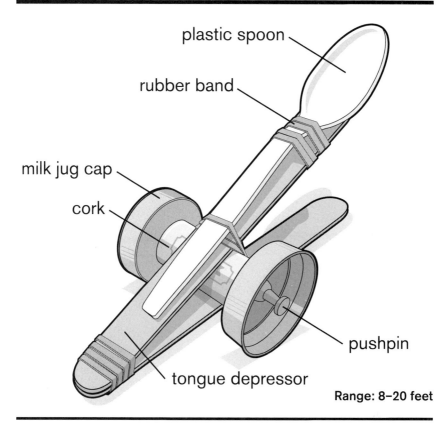

plastic spoon

rubber band

milk jug cap

cork

pushpin

tongue depressor

Range: 8–20 feet

Don't let this unassuming little device fool you. While it takes only seconds to construct, the Depressor Spoon Catapult is designed to destroy. Its straightforward design is perfect for mass production and outfitting an army. Plus, no tools are required!

Supplies

2 tongue depressors
3 rubber bands
1 cork
1 plastic spoon
2 plastic milk jug caps
    (optional)
2 pushpins (optional)

Tools

Safety glasses
Hot glue gun (optional)

Ammo

1+ mini marshmallows

Step 1

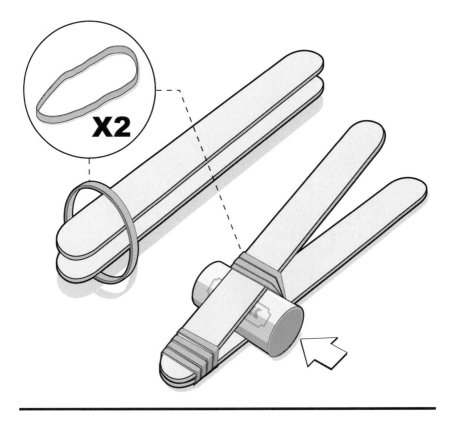

Bundle two tongue depressors at the end with one rubber band. Then, slide a cork between the two tongue depressors as close to the attached rubber band as possible. The tension of the tongue depressors will force the cork out, so wrap another rubber band around both tongue depressors behind the cork to prevent this from happening.

Step 2

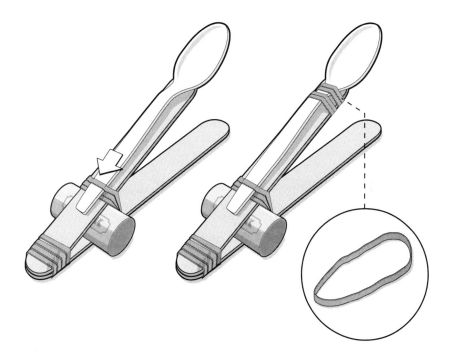

Slide a plastic spoon under the rubber band located behind the cork. Then add an additional rubber band around the neck of the spoon, securing it to the top of the tongue depressor.

Test the catapult by placing one hand on the bottom depressor to stabilize it, and then pull down the top arm to launch.

Step 3

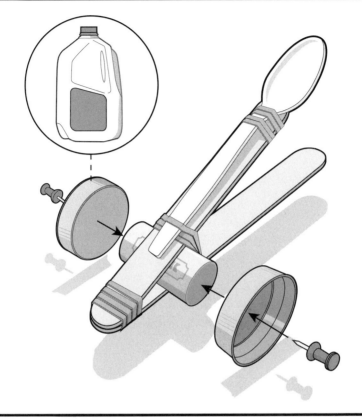

To add wheels (optional), center two pushpins through two plastic milk jug caps, and then attach both wheels to the cork by inserting the pushpins through the caps and into the cork. To reinforce the connection, carefully add hot glue to the back of the wheels if needed. The wheels don't need to be added if supplies are limited or you're concerned about safely using pushpins with a young crowd.

Remember to use eye protection! This catapult excels at height because of its angle, so never lean over this MiniWeapon when firing. *Never aim this catapult at a living target and use only safe ammunition.* Mini marshmallows work nicely.

MARSHMALLOW CATAPULT

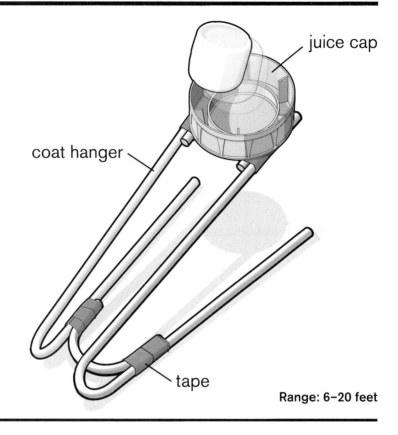

juice cap

coat hanger

tape

Range: 6–20 feet

Yes, this catapult is designed to launch large, puffy marshmallows—how awesome! It's made from two plastic coat hangers and a large plastic juice jug cap, so assembly is lickety-split. Armed with soft ammo, this lighthearted siege weapon is the perfect choice for indoor use. Pair it with the Attacking Army target found in the back of the book (page 259) for practice.

Supplies

2 plastic coat hangers
Duct tape
1 plastic juice jug cap (or similar)

Tools

Safety glasses
Pliers or wire cutters
Hot glue gun

Ammo

1+ large marshmallows

Step 1

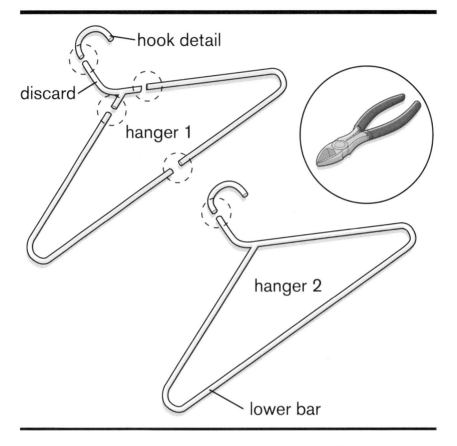

hook detail

discard

hanger 1

hanger 2

lower bar

With pliers or wire cutters, remove the hook detail from two plastic hangers as illustrated.

Cut hanger 1 down the middle by snipping the center of the lower bar and making two additional cuts on each side of the neck detail. Discard the neck detail from hanger 1.

Only the hook detail will be used from hanger 2. The remaining material from hanger 2 can be saved and used for either the Plastic Hanger Bow (page 119) or the Wooden Ruler Crossbow (page 143).

Step 2

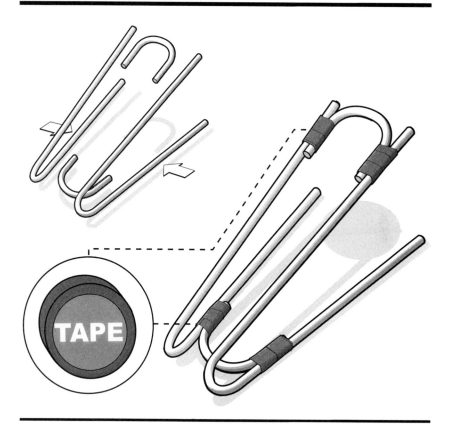

TAPE

The hook pieces will become the braces to hold both hanger halves together to make the frame. To do this, tape the first hook detail onto the base, near the front. The second hook detail should be taped to the top rear frame, as shown.

Step 3

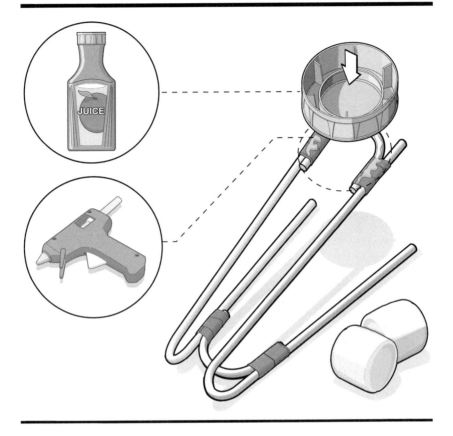

The basket for this catapult is made from a plastic juice jug cap because it has a large, 2½-inch diameter. Wash and dry the selected cap. Although other caps similar in size can be substituted, **do not use a cap from hazardous materials**, especially if you plan on launching anything edible. Carefully hot glue the cap with the smooth surface fixed to the upper hook detail on the frame and allow the glue to dry.

Remember to use eye protection! To fire the catapult, support the lower frame with one hand while you load and pull back the swing arm with the other. It's important that you accompany the launch with your best battle cry!

Marshmallows are soft, but ammunition can be substituted without notice! **Never aim this catapult at a living target and use only safe ammunition.**

MOUSETRAP CATAPULT

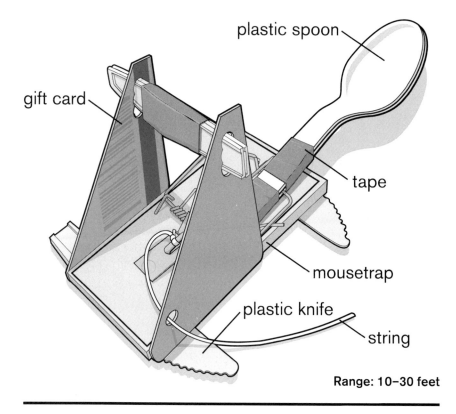

plastic spoon

gift card

tape

mousetrap

plastic knife

string

Range: 10–30 feet

Mounted with a ready-made spring-loaded mechanism, the Mousetrap Catapult is one of the most powerful catapults employed in any Middle Ages mini-battle! Integrated with a built-in trip release, it will angrily propel any projectile on command. Your enemy will walk right into this trap!

Supplies

1 mousetrap (unused)
2 plastic spoons
1 expired or zero-balance
    plastic gift card
2 plastic knives
Duct tape
String

Tools

Safety glasses
Hot glue gun
Scissors
Single-hole punch

Ammo

1+ soft candies or pencil
    erasers

Step 1

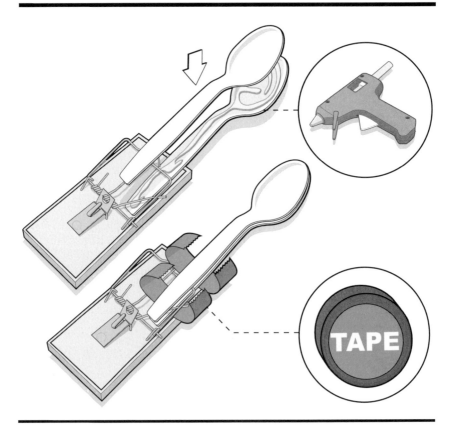

With the mousetrap held tightly in the set position to prevent it from snapping shut, place one plastic spoon beneath the mousetrap bar—but off center. Then carefully hot glue a second spoon directly on top, sandwiching the mousetrap bar. Hold this assembly together as it dries. Then wrap tape around both spoons to reinforce the catapult swing arm.

Step 2

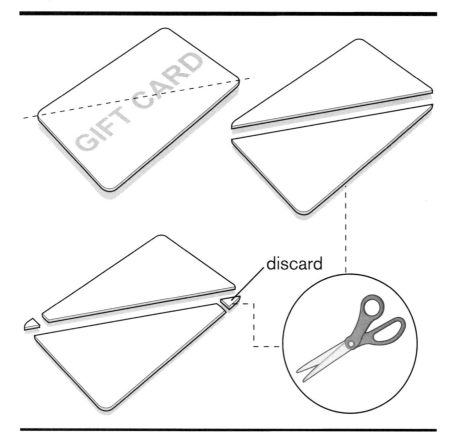

discard

Use scissors to cut a diagonal line across an old gift card, from its top right corner to its bottom left corner, making two triangles (top images).

As a safety precaution, cut roughly ¼ inch off the pointed tips on the two triangles (bottom image). Discard these tips.

Step 3

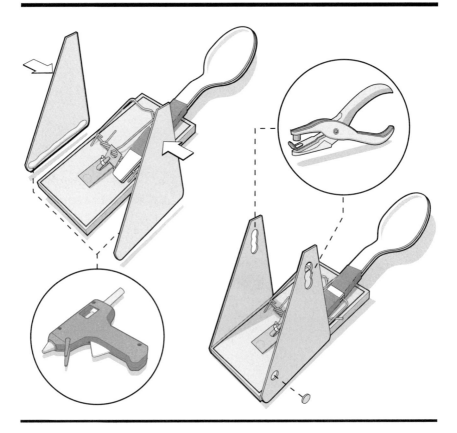

Carefully hot glue both card triangles onto the sides of the mousetrap frame. Both triangles should be angled toward the front and flush with the front and base of the mousetrap frame.

Next, cut holes into each card triangle using a single-hole punch. The first hole should be near the top of the card triangle. After the first hole has been punched, create one large opening by punching two more holes overlapping the first (right image). Repeat on the other card.

The last hole should be punched in the lower corner of one card, in front of the trip release.

Note: The illustration above depicts the mousetrap in the set position, but it does not need to be in the set position in this step.

Step 4

Position two plastic knives along the mousetrap frame as shown and use scissors to cut the knife handles 1 inch longer than the width of the frame. Do not discard the plastic knife ends.

Note: The illustration depicts the mousetrap in the set position, but it does not need to be in the set position in this step.

Step 5

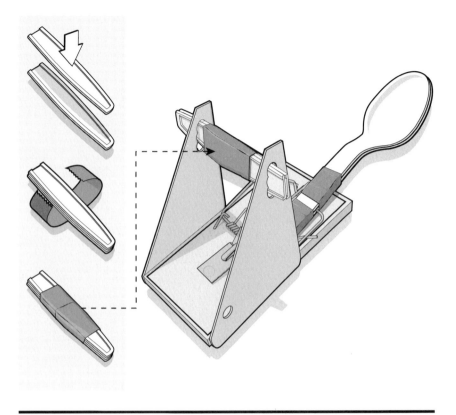

Bundle the cut handles from the plastic knives as shown. Secure the assembly with tape.

Slide the bundle into the two larger holes you created at the tips of the card triangles. A snug fit is best to prevent the bundle from dislodging when struck by the swing arm. If the bundle does not fit, adjust the hole(s) with the hole punch.

Step 6

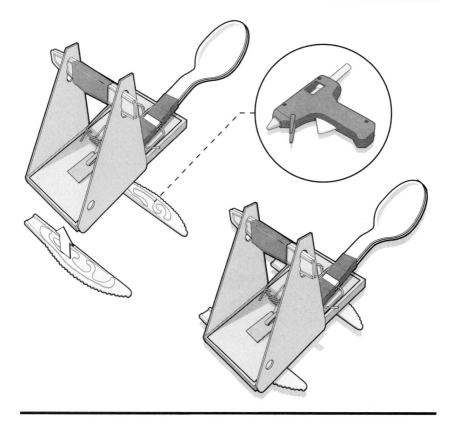

Base supports must be added to the frame to balance the catapult during firing. On opposite ends of the underside of the mousetrap frame, carefully hot glue both plastic knife ends (from step 4) to add support.

Step 7

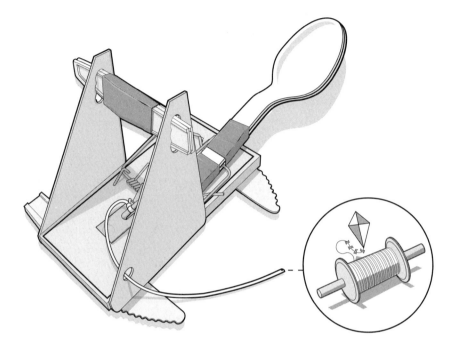

For the last step, knot at least 6 inches of string to the trip release. Once it is securely fastened, feed the string through the lower punched hole in the plastic frame.

To fire the Mousetrap Catapult, carefully set the mousetrap and then load ammo into the plastic spoon. With one hand, hold down the frame on either side, and use your other hand to pull the string. If the power of the catapult pulls it away from your hand, use rubber bands to mount the catapult to a heavy book.

Remember to use eye protection, especially for a spring-loaded launcher! Since the Mousetrap Catapult is spring-loaded, anything can happen. ***Never aim this catapult at a living target and use only safe ammunition.***

CD TREBUCHET

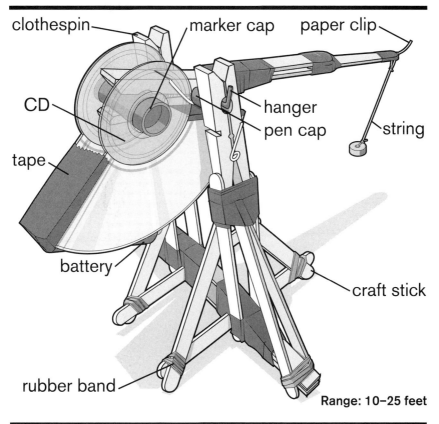

clothespin

marker cap

paper clip

CD

hanger

pen cap

string

tape

battery

craft stick

rubber band

Range: 10–25 feet

One of the most powerful catapults of the Dark Ages, the trebuchet could hurl hundred-pound projectiles into enemy fortifications with devastating impact. In both function and silhouette, the CD Trebuchet is the perfect miniature representation of this intimidating machine!

Supplies

19 craft sticks
10 rubber bands
5 clothespins
Duct tape
1 small paper clip
2 unwanted CDs or DVDs
5 AA batteries (noncorrosive)
1 standard art marker with cap
1 plastic ballpoint pen cap
1 metal coat hanger

Heavy thread or kite string

Tools

Safety glasses
Large scissors or hobby knife
Hot glue gun
Pliers or wire cutters

Ammo

1+ ring-shaped candies

Step 1

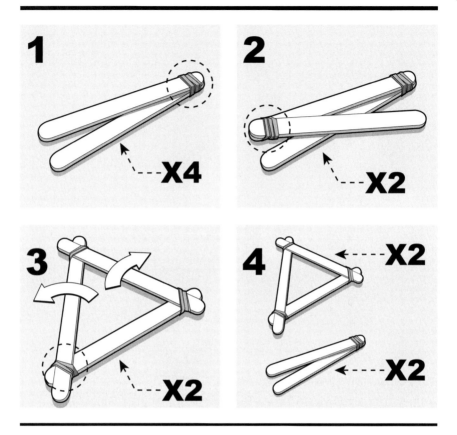

Divide eight craft sticks into four stacks; each stack will be two sticks high. Fasten each stack using one rubber band at the end (1). Set two of the completed bundles aside; you will use them in step 4 (page 91).

For the remaining two bundles, rubber band an additional craft stick to the end of one of the attached craft sticks (2).

To construct the triangular frame, secure the two loose-ended craft sticks to one another using another rubber band (3).

When finished, you should have two triangular craft stick assemblies and two double craft stick assemblies (4).

Step 2

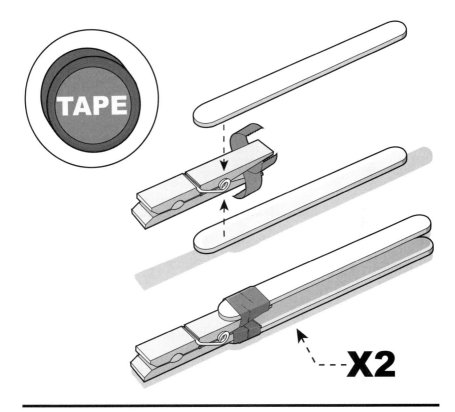

You will now build two swing arm supports for the trebuchet, using four craft sticks and two clothespins. Tape one craft stick approximately 1 inch onto the back of each clothespin prong as illustrated. Repeat this step until you have two completed assemblies, as shown at the bottom.

Step 3

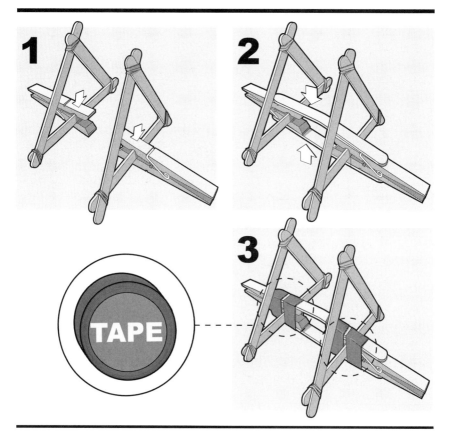

To start the trebuchet frame, clamp one clothespin onto each triangle assembly. When attached, the clothespin should be centered and fixed at a 90-degree angle. Repeat this step for both triangular frames. The finished assembly should stand upright (1).

Mirror the assemblies approximately 2½ inches apart from one another, then place a craft stick on the top and bottom of the attached clothespins (2).

To support the assembly, tightly tape it together with four pieces of tape; see the illustration for placement (3).

Step 4

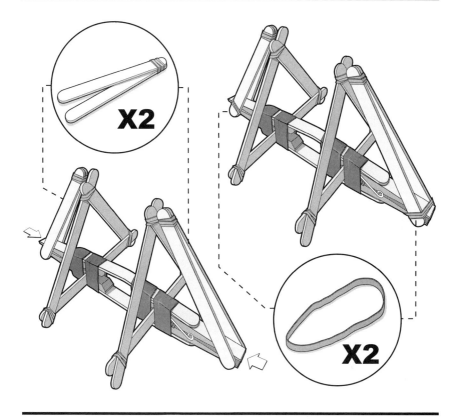

The two bundles you set aside in step 1 are the trebuchet frame supports. Attach these bundles by sliding the non-rubber-banded side around the ends of the attached clothespins, as pictured on the left. Angle the bundles into the triangle frame as shown.

Once they're in place, use one rubber band on each side to secure the craft sticks to the clothespins. Additional supports to secure the frame will be added in the next steps.

Step 5

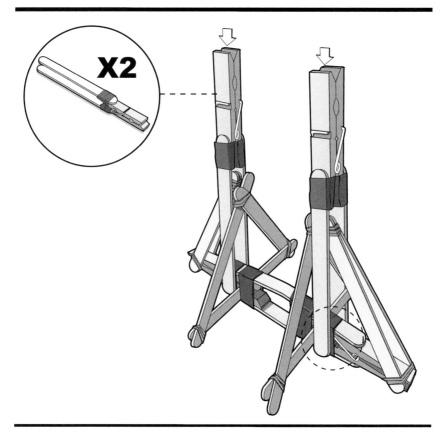

X2

Next, slide both swing arm supports (from step 2) upright onto the frame assembly. They should be placed outside the triangle support, sandwiching the lower frame as shown. Also, at this point the 45-degree braces you added in step 4 should be resting under both swing arm supports, between the taped clothespin prongs.

Step 6

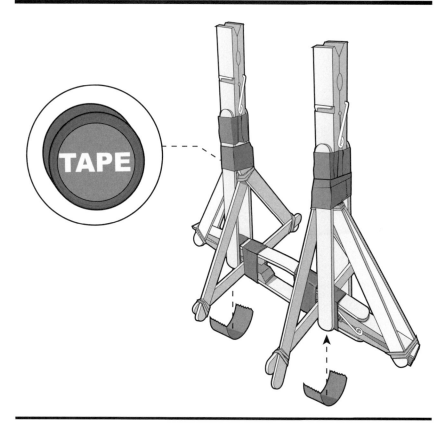

Align both swing arm assemblies at 90-degree angles, then secure the frame by tightly wrapping tape around the swing arm supports, triangle, and angled braces as shown.

Then add tape to the lower assembly. This will prevent the frame from shifting during firing. With the added tape, the trebuchet frame should be solid. If not, add more tape or supports.

Step 7

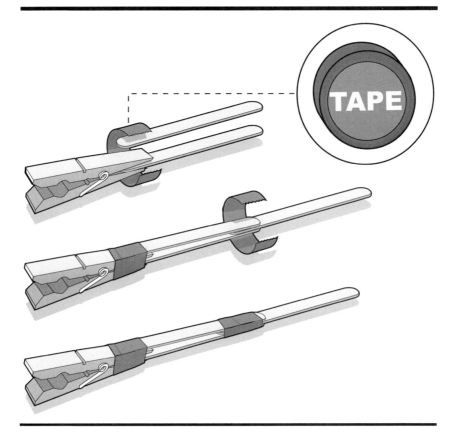

Set the trebuchet frame assembly aside; it's time to construct the trebuchet's swing arm. This step requires one clothespin and three craft sticks.

Place the first craft stick between the rear prongs of the clothespin, inserted roughly 1 inch deep. Then place another craft stick directly on top of the prong, lined up with the first craft stick. Use tape to tightly wrap both craft sticks in place (top image).

Next, slide the last craft stick between the ends of the double craft stick assembly, with roughly 1 inch of overlap. Then secure all three sticks with tape.

Step 8

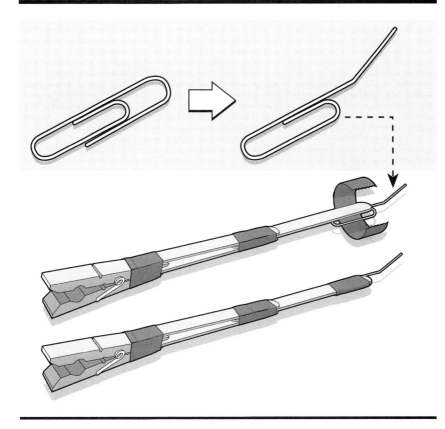

To complete the swing arm assembly, add a hook release. It will hold and launch the projectiles. Straighten one end of the paper clip as illustrated, with a slight upward bend.

Next, tape the modified clip to the end of the swing arm assembly, with the straightened end of the clip extending past the craft sticks. To adjust the range of the trebuchet, you can make additional adjustments to the clip angle once it is complete.

Step 9

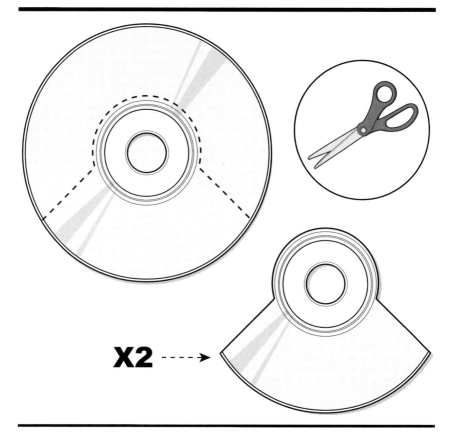

X2 - - - - ▶

What is unique about this siege weapon is that it's powered by gravity, unlike many of the torsion-powered catapults. During battle, the raised counterweight is released, rotating the attached swing arm and launching the projectile toward the target.

Start the construction of the counterweight frame with two unwanted CDs or DVDs. With large scissors or a hobby knife, carefully cut both discs using the illustration as a guide. When you're finished, both CDs should look like the bottom image. Note: Cutting a CD with scissors is delicate work. If you rush, the CD may crack. It is best to make several cuts and slowly remove the material instead of trying to make a single quick cut.

Step 10

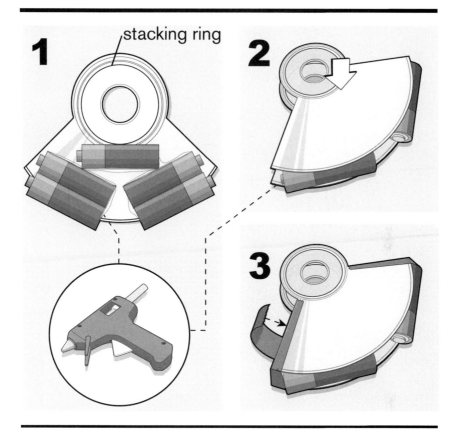

It's time to add the counterweight! Carefully hot glue five or more used AA batteries (or similar) to one of the modified CDs. The batteries need to stay outside the center circle, also known as the stacking ring or plastic hub area (1).

Once the glue has dried, hot glue the second modified CD on top of the attached batteries, aligning the perimeter and center of the circle (2). For aesthetic reasons, tape the side of the disc detail to hide the batteries (3).

AA batteries are recommended because their width is similar to that of a clothespin. However, *if the AA batteries are damaged, do not use them!* Instead, other batteries, coins, or another weighted mate-rial can be substituted. Added weight will produce different results, so experimentation is suggested for this project.

Step 11

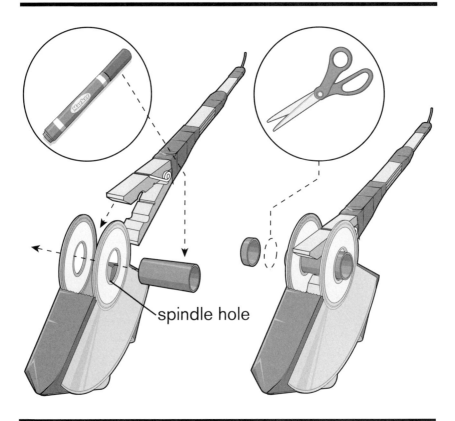

spindle hole

With the counterweight completed in the last step, you will now attach the swing arm assembly using the cap from a standard marker. First, prefit the marker cap through the counterweight assembly (spindle hole). The cap should be able to rotate freely. If it can, shorten the cap to be slightly wider than the counterweight assembly by carefully cutting it with the large scissors or hobby knife.

Now, with the marker cap in the counterweight assembly, attach the swing arm by snapping the attached clothespin between the CDs and onto the marker cap. If properly assembled, the swing arm should rotate back and forth unobstructed.

Step 12

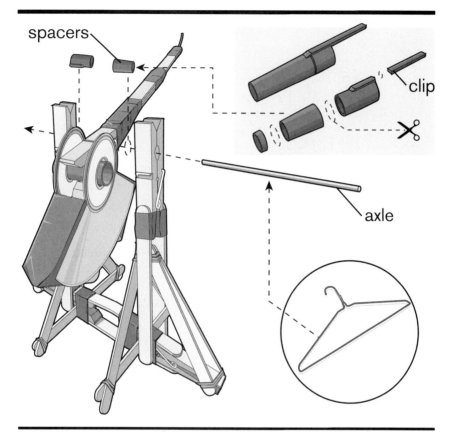

spacers

clip

axle

In this step you will prepare two different components that will be used to hold and center the swing arm assembly onto the frame. Prior to starting, locate one plastic ballpoint pen cap and one metal hanger.

With pliers or wire cutters, cut a straight section out of the metal coat hanger approximately 1½ inches longer than the distance between the frame supports. This rod will be the swing arm axle.

Next, with the large scissors or hobby knife, remove the clip and enclosed tip from the plastic pen cap (top right image). Then cut the cap in half, creating two equal-length cylinders; these will become the swing arm spacers.

Slide the coat hanger axle through the rounded gap on the first clothespin mounted on the frame. Then add one pen cap spacer to the axle, and slide the axle through the spring detail on the swing arm clothespin, through the second spacer, and finally through the gap on the other clothespin mounted on the frame.

Step 13

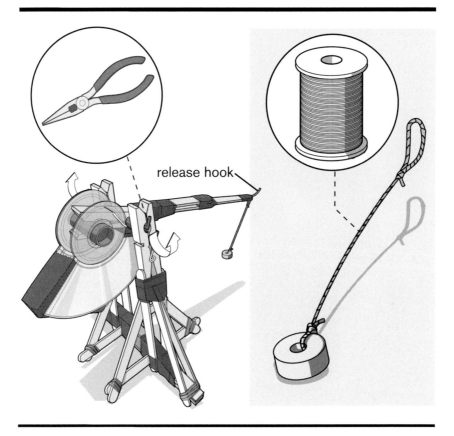

release hook

With pliers or wire cutters, make two 90-degree bends at the end of the hanger axle, locking it in place. The swing arm should now easily swing back and forth (left illustration).

Finally, it's time to construct the ammunition. Because these projectiles really fly, it might be worthwhile to manufacture several prior to attacking. Start with a 5-inch piece of heavy thread or kite string. Tie one end into a loop and then tie the opposite end through a ring-shaped candy (or around an eraser pulled from a pencil).

To fire, pull down the swing arm and place the string loop over the hook release on the swing arm. Once attached, place the ammo beneath the framework. Hold the frame and release the arm. Adjust the release hook angle for different results.

Consult the safety instructions on page ix! The swing arm has a small hook release (paper clip) that can rotate unexpectedly and then swing back and forth.

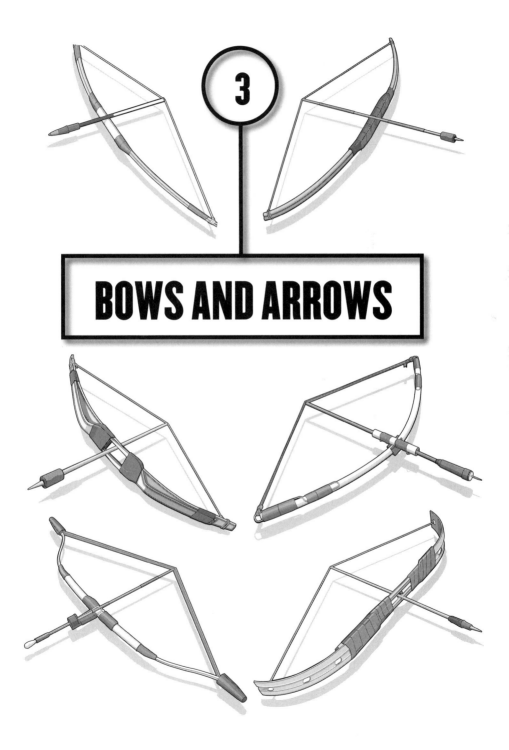

BOWS AND ARROWS

DOUBLE SKEWER BOW

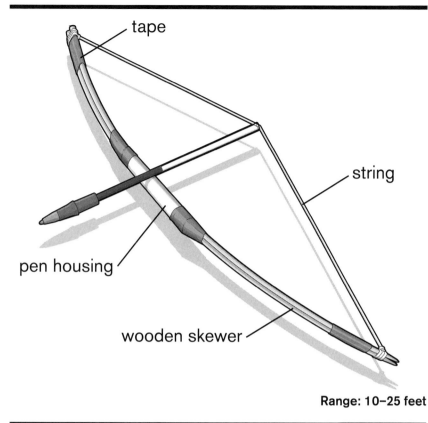

tape

string

pen housing

wooden skewer

Range: 10–25 feet

To become a master archer, a beginner bowman must first study the fundamentals of archery. This Double Skewer Bow is the perfect tool for any peasant who wants to become skilled in the use of the bow and arrow. Constructed from everyday materials, this double-reinforced wooden straight bow can quickly be assembled for practice.

Supplies

1 plastic ballpoint pen
2 wooden skewers
Duct tape
String

Tools

Safety glasses
Large scissors or hobby knife

Ammo

1+ pen ink cartridges

Step 1

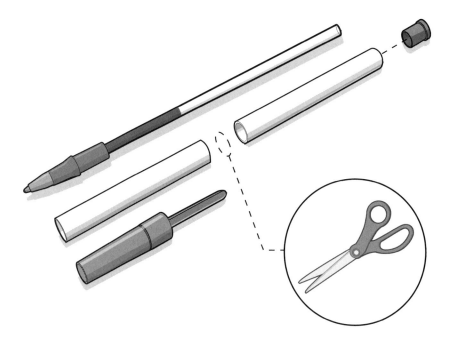

Pillage the kingdom for one plastic ballpoint pen. Disassemble the pen into its various parts. Depending on the pen, a tool may be needed to dislodge the rear pen-housing cap. Large scissors, a hobby knife, or small pliers should do the job.

Next, use large scissors or a hobby knife to cut the pen housing in half. You will only use one half of the housing in this step, but you can save the other half and the pen cap for building the Plastic Hanger Bow (page 119). The ink cartridge will be used as an arrow for this project.

Step 2

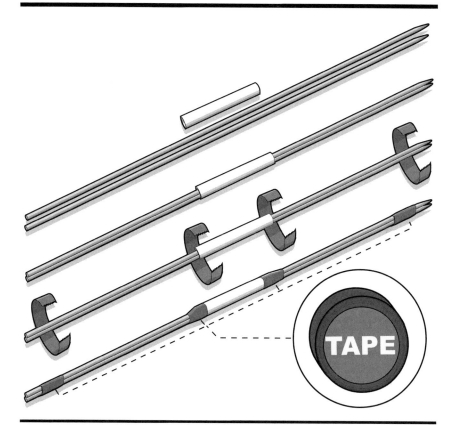

Slide the half of the pen housing over two wooden skewers until it is centered on the skewers, then tape both ends of the pen housing into place. Apply two additional pieces of tape to the ends of the skewers, approximately ¼ to ½ inch from the tips.

Step 3

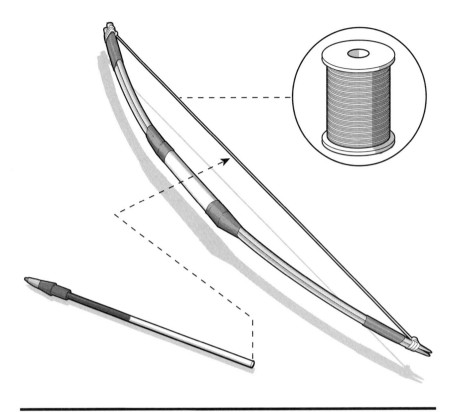

Tie the tip of the bow assembly with string. Knot the string several times, using the wedge-shaped gap between the two skewers to keep the string in place. Slightly bend the bow and then tie the opposite end with several knots. Trim off any extra string.

The Double Skewer Bow is complete—time to test fire! Center the ink cartridge arrow onto the pen handle, as shown on page 103, then grasp both the string and arrow slowly. Once the bow is drawn, aim and release the bowstring. *If you're concerned that the ink cartridge might explode on impact*, replace this arrow with one of the many other options found in this book.

Remember to use eye protection! Arrows can travel at a high velocity and have sharp points. *MiniWeapon projects are not meant for use on living targets*. Always stay clear of spectators and fire the bow in a controlled manner. Homemade weaponry can malfunction.

CHOPSTICK BOW

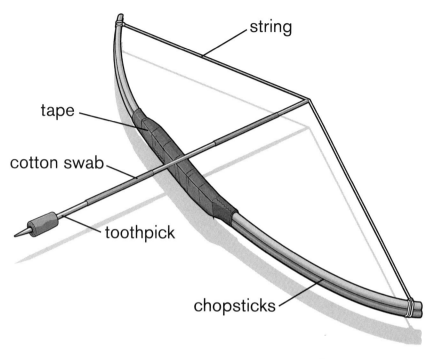

string

tape

cotton swab

toothpick

chopsticks

Range: 10–25 feet

Similar in design to the Yumi bow used by samurai warriors in Japan, the Chopstick Bow is fashioned from traditional wooden chopsticks, making the core stiff. Its range is measured by its user's strength. Master its powerful draw weight while practicing *kyudo*–Japanese archery.

Supplies

2 sets of wooden chopsticks
Duct tape
String

Tools

Safety glasses

Bowl of warm water (optional)
Scissors

Ammo

2+ plastic cotton swabs
3+ toothpicks
Clear tape

Step 1

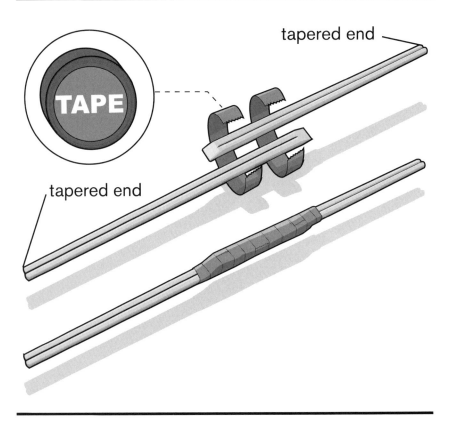

tapered end

TAPE

tapered end

Overlap the nontapered ends of two sets of chopsticks. The overlapping area should be approximately 3 inches long. Once the chopsticks are aligned, tightly tape them together to form the bow.

This bow will be straight. However, it is possible to make a curved bow by bending the chopsticks prior to assembly. To do so, place the wooden chopsticks in a bowl of warm water for 60 minutes or more to soften the wood. Remove the chopsticks from the water, slowly bend them into an arc, and then let the sticks dry to maintain the arc form. The chopsticks must be completely dry before taping.

Step 2

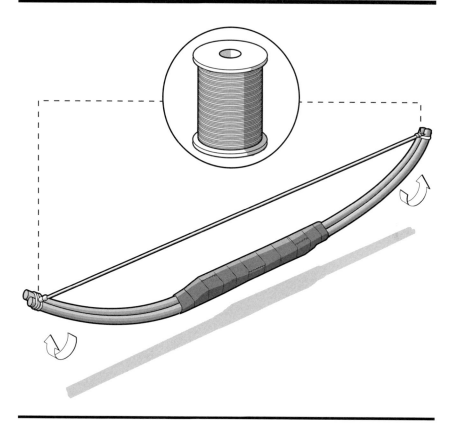

At one end of the chopstick frame, tie the string with a single or double knot, placing the string between the chopsticks to hold it in place. Then slowly bend the bow, and hold the form while the opposite end is tied with the same strength.

Note: Depending on how the chopsticks are manufactured, the wood density and strength will vary, so excessive bending can result in breakage.

Step 3

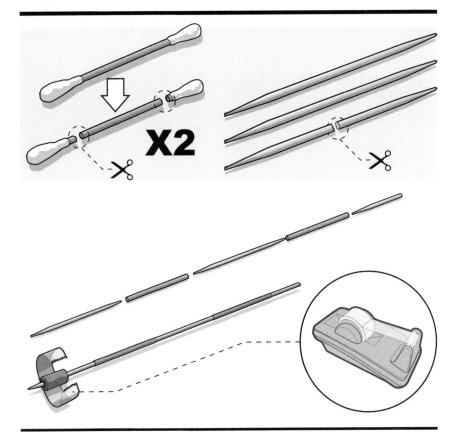

Each arrow will be constructed from two plastic cotton swabs and three toothpicks.

    Cut the two ends off both cotton swabs, as shown. Then cut one of the three toothpicks in half, as shown. Slide a single round toothpick into the plastic cavity of the cotton swab, then repeat this step by joining the additional toothpicks and cotton swabs together in this order: toothpick–cotton swab–toothpick–cotton swab–half of toothpick. The assembly will make up one arrow; the half toothpick will be the nock (end).

    To increase the arrow's accuracy, add weight at the point. Tightly wind a 12-inch section of clear tape around the toothpick about ½ inch from the point. You may need to decrease or increase the amount of tape until you arrive at the proper balance. You can also substitute arrows used in the other projects in this book.

    Consult the safety instructions on page ix! Arrows can travel at a high velocity and have sharp points.

PLASTICWARE BOW

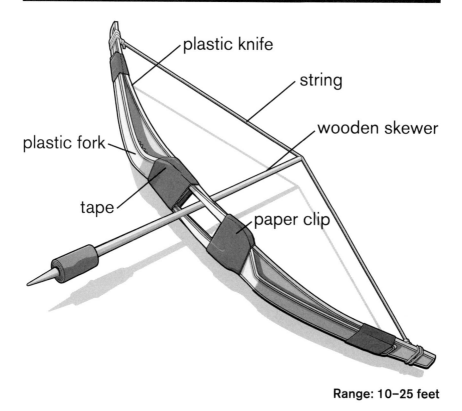

plastic knife

string

wooden skewer

plastic fork

tape

paper clip

Range: 10–25 feet

Made from plastic utensils, the Plasticware Bow will certainly raise a few eyebrows in the king's court. Even the juggling jester will take notice. Its unique frame, which slightly resembles a recurve bow, adds extra tension, giving the fired arrow some extra zing!

Supplies

2 plastic forks
Duct tape
2 large paper clips
2 plastic knives
String

Tools

Safety glasses

Small pliers or wire cutters
(optional)
Power drill (optional)
Scissors

Ammo

1+ wooden skewers
Clear tape

Step 1

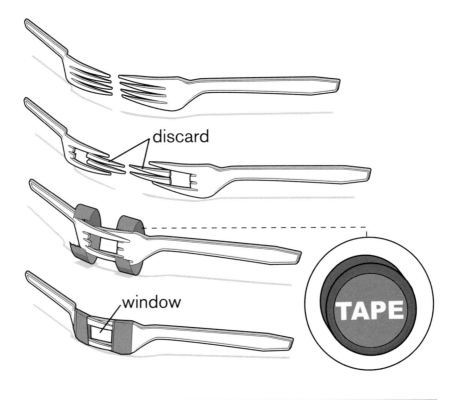

discard

window

TAPE

Start with two plastic forks. With your fingers, snap off the two center prongs on both forks. Depending on the plastic's thickness, you may need small pliers or wire cutters to remove the prongs. Discard the removed prongs.

Position the forks with the prongs facing one another, overlapping the remaining prongs to create a small window. Join both forks by tightly wrapping tape around the fork housing, but do not cover the opening with tape.

Step 2

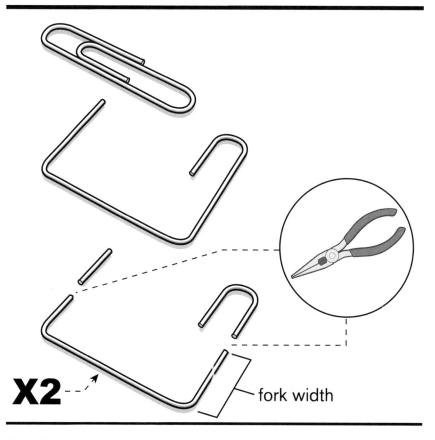

X2

fork width

Next, bend the opposite sides of a large paper clip to make two 90-degree angles, creating a large "U" frame. Use pliers or wire cutters to trim the paper clip's width to match the plastic forks' width. Repeat this step so that you end up with two "U" frames.

Step 3

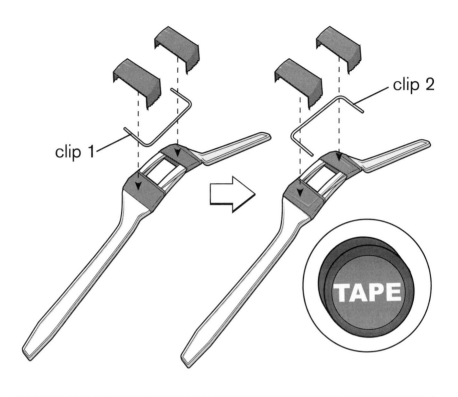

Tape the first paper clip "U" frame to the underside of the forks. The bent paper clip should fit around the opening of the assembled forks.

Next, tape the second paper clip frame to the underside of the forks, mirroring the previously attached paper clip. It is important that this connection is tight.

Step 4

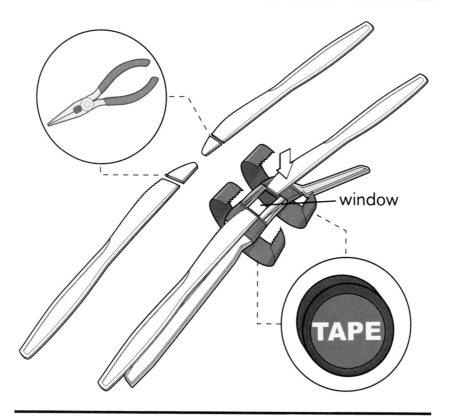

window

TAPE

With scissors or wire cutters, carefully remove ¼ inch from the tips of both plastic knives, as shown.

Next, align both knives to the underside of the fork assembly, then tightly tape both the knife ends onto the forks. Do not cover up the window.

Step 5

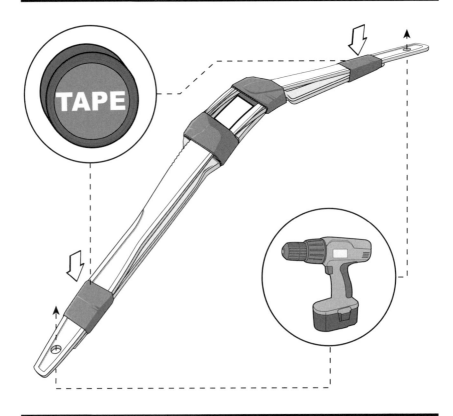

Further secure the assembly by tightly wrapping tape around the handle of both the fork and the knife, on both sides, as illustrated.

On opposite ends of the bow assembly, approximately ½ inch from the tip of the fork handle, drill two string-sized diameter holes. The plastic might crack, so drill slowly and use a new drill bit. No drill? No problem—the string can still be attached to the limbs in step 6 using tape only (not shown).

Step 6

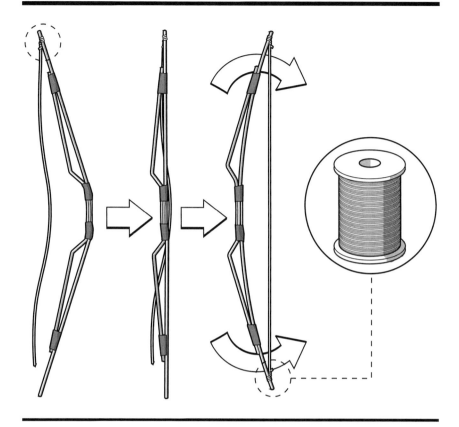

At one end of the plastic frame assembly, tie the string with a double or single knot using the drilled hole to secure the string in place.

You will now add tension to the bow by bending the assembly backward as illustrated. Once the frame is pulled back, tie the opposite end with the attached string; several knots are recommended. Carefully remove any excess string with scissors. The Plasticware Bow is complete.

Step 7

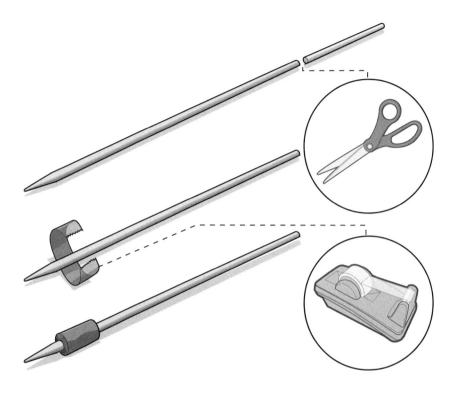

The arrow will be constructed from one wooden skewer. Cut the skewer with scissors so that its length is roughly 8 inches long.

To increase the arrow's accuracy, add weight to the point by tightly winding a 12-inch section of clear tape around the skewer about ½ inch from its point. You may need to decrease or increase the amount of tape until you arrive at the proper balance. If you don't have wooden skewers, try the arrows used in the other bow and crossbow projects.

To fire, center the arrow in the middle of the bow frame, between the removed prongs, then grasp both the string and arrow and pull back slowly. Once the bow is drawn, aim and release the bowstring to send the arrow flying. ***Remember to use eye protection!*** Arrows can travel at a high velocity and have sharp points. ***MiniWeapon projects are not meant for living targets***. Always stay clear of spectators and operate the bow in a controlled manner. Homemade weaponry can malfunction.

PLASTIC HANGER BOW

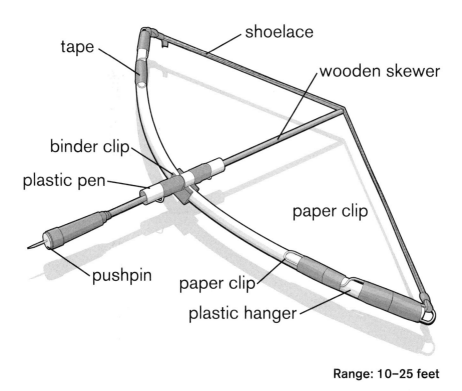

tape

shoelace

wooden skewer

binder clip

plastic pen

paper clip

pushpin

paper clip

plastic hanger

Range: 10–25 feet

It's called the Plastic Hanger Bow and it's the perfect peasant weapon, because it uses the most basic materials but fires heavy-gauge, steel-tipped arrows capable of penetrating almost any tin can armor.

Supplies

1 plastic hanger
2 large paper clips
1 plastic ballpoint pen
Duct tape
1 small binder clip (19 mm)
Shoelace
1 pushpin

Tools

Safety glasses
Wire cutters
Large scissors or hobby knife
Hot glue gun

Ammo

1+ wooden skewers
1+ pushpins
1+ plastic pen caps
Clear tape

Step 1

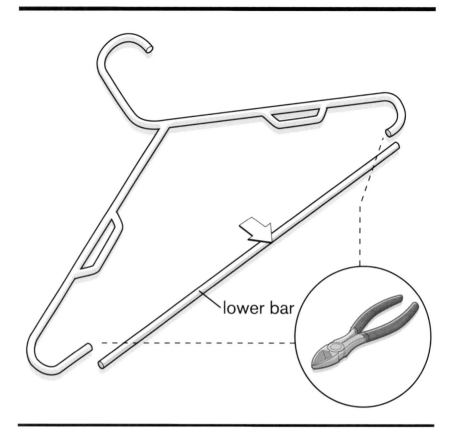

lower bar

Remove the lower bar from one plastic coat hanger using wire cutters. You will only use the removed section; the rest of the hanger should be recycled.

Step 2

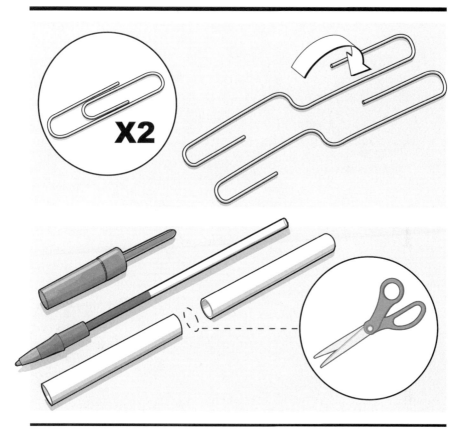

Bend two large paper clips flat (180 degrees) as shown, making each look like the letter "S."

Next, disassemble one plastic ballpoint pen into its various parts. Once the ink cartridge is removed, carefully cut the pen housing in half with large scissors or a hobby knife. You will not be using the ink cartridge for this MiniWeapon; save it for additional bolts for the Pen Ballista (page 191).

Step 3

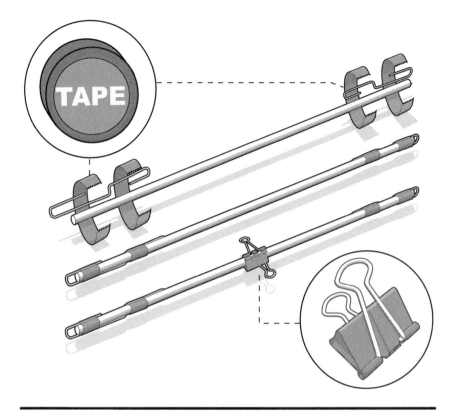

Tape the two modified paper clips to the opposite ends of the plastic hanger's lower bar. Both paper clip loops should protrude ¼ inch from the lower bar. Due to the stress these clips will endure, add additional tape to them to prevent malfunction during use. The illustration suggests placement of added tape.

Next, clamp one small binder clip to the center of the lower bar. This clip is the beginning of the attached arrow guide.

Step 4

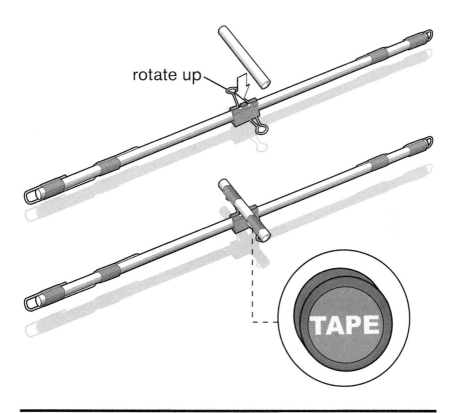

rotate up

TAPE

Now center half of the pen housing onto the two flipped up metal handles attached to the small binder clip. Once in place, tape both sides, with the two metal handles laying flush across the pen housing.

Step 5

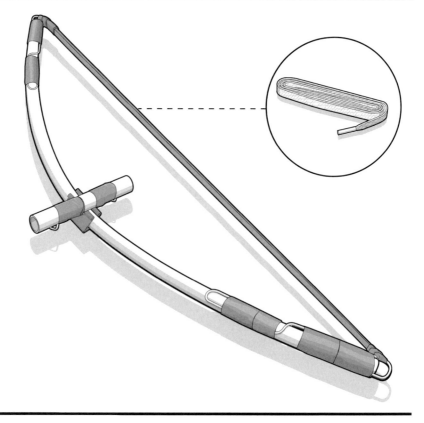

Attach a shoelace to one end of the bow, using the fastened paper clip loop as the tie-down. A double knot is suggested. (A shoelace bowstring is recommended; however, string or a rubber band can be substituted.)

Next, make a slight bend in the bow and then tie the shoelace to the opposite end of the bow, making sure the bowstring is tight. Cut away any excess shoelace at each end.

Step 6

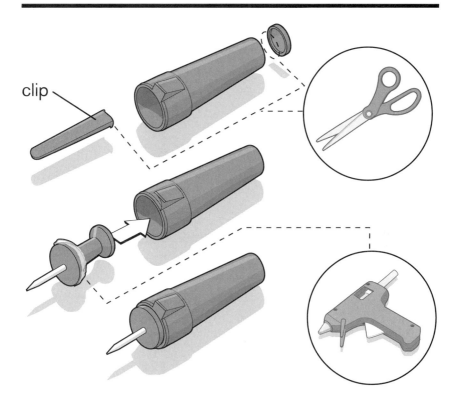

clip

Remove both the clip and the tip from a plastic pen cap using the large scissors or hobby knife, as shown. Discard the clip and tip.

Dab a small amount of hot glue inside the pen cap and then slide the pushpin (point out) into the pen cap opening. Depending on the diameter of both the pen cap and the pushpin, hot glue may not be needed. If hot glue is unavailable, wrap tape around the pushpin handle to create a snug fit.

Step 7

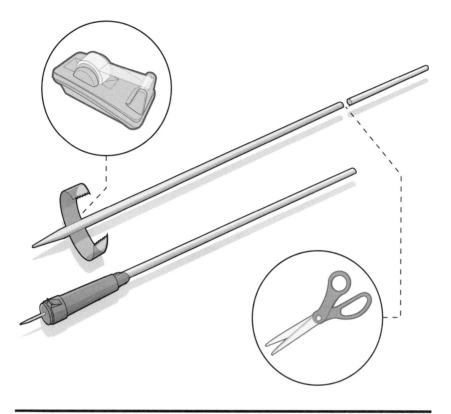

The shaft of the arrow is made from a wooden skewer. Carefully cut the skewer so that its length is roughly 8 inches. (If a wooden skewer is unavailable for this project, refer to the arrows used in the other bow and crossbow projects in this book.) To add the point, tightly wrap 5 inches of clear tape around the skewer's shaft, approximately ½ inch from the tip. This will increase the skewer's diameter.

Slide the pointed pen cap over the taped end of the skewer for a snug fit. Add or reduce tape if needed; hot glue can also be used. The pointed arrow is complete!

When firing, use the arrow guide mounted on the bow to increase accuracy. Pull and watch the arrow fly! **Remember to use eye protection!** Arrows can travel at a high velocity and have sharp points. **MiniWeapon projects are not meant for living targets**. Always stay clear of spectators and operate the bow in a controlled manner. Homemade weaponry can malfunction.

ADVANCED PEN BOW

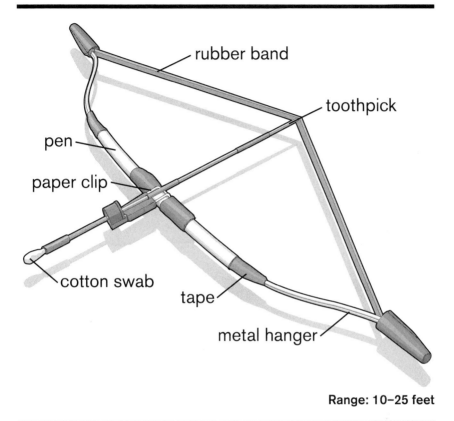

rubber band

toothpick

pen

paper clip

cotton swab

tape

metal hanger

Range: 10–25 feet

Renowned for its military precision, the Advanced Pen Bow lends sophistication to any homemade armory. Perfect for competitive shooting, this bow is equipped with a frame-mounted arrow guide, increasing its accuracy. Launch countless arrows from its indestructible composite frame and destroy targets at will.

Supplies

2 metal coat hangers
2 plastic ballpoint pens
2 large paper clips
Duct tape
1 large rubber band

Tools

Safety glasses

Wire cutters
Pliers (optional)
Large scissors
Hot glue gun

Ammo

3+ toothpicks
2+ plastic cotton swabs
Clear tape

Step 1

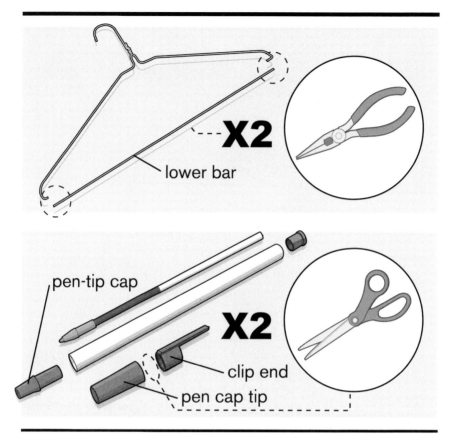

Remove the lower bar from two metal coat hangers using wire cutters. Both lower bars will be used for this project, but the remaining metal material should be recycled.

Next, disassemble two plastic ballpoint pens into their various parts. Depending on the pens, you may need a tool—pliers or the lower bar—to dislodge the rear pen-housing cap.

With large scissors, cut both pen caps approximately ¼ inch from the clip end.

Step 2

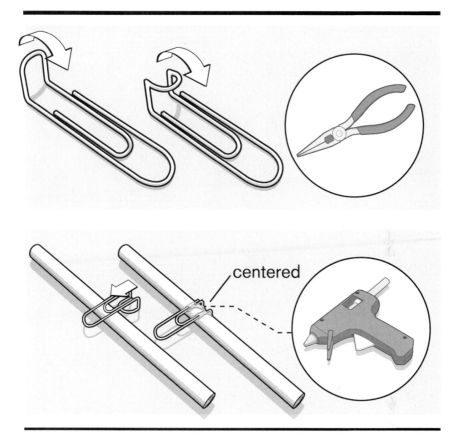

centered

With wire cutters or pliers, make one 90-degree bend ⁵/₁₆ inch (the diameter of the pen housing) from the end of a large paper clip. Then make another 90-degree bend ⁵/₁₆ inch from the previous bend. Only modify one of the two paper clips; the other paper clip should remain straight.

Center the double 90-degree bent paper clip on the pen housing as shown. Carefully hot glue this paper clip in place.

Step 3

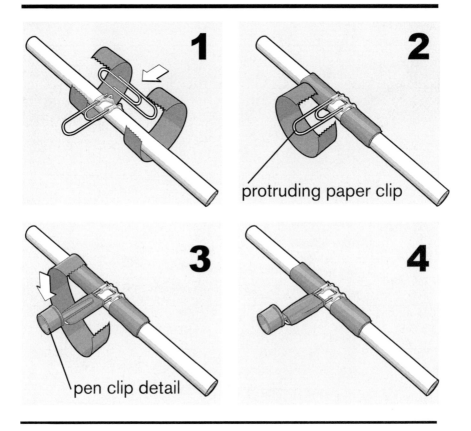

1

2

protruding paper clip

3

pen clip detail

4

To help support the glued paper clip on the pen housing, the second paper clip (straight) should lie across the 90-degree bend behind the pen housing. Secure the straight paper clip onto the pen housing using two pieces of tape (1).

Add one 2-inch piece of tape around the protruding paper clip you glued on in step 2 (2). Next, rest the pen clip detail on top of the protruding paper clip, and then tape only the clip into place (3). This completes the arrow guide (4).

Step 4

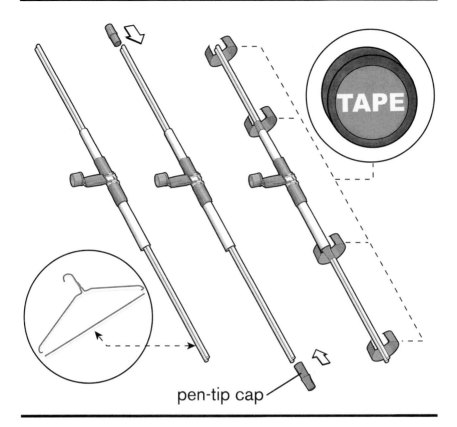

pen-tip cap

Slide the pen assembly onto both lower hanger bars cut in step 1. Then, on opposite ends, slide two pen-tip caps onto the bars. Snap them into the pen housing frame, as shown on the right.

Center the pen assembly onto the lower bar and tape it into place by adding tape around the ends of the pen housing. Then add more tape ¼ inch from the end of the bars.

Step 5

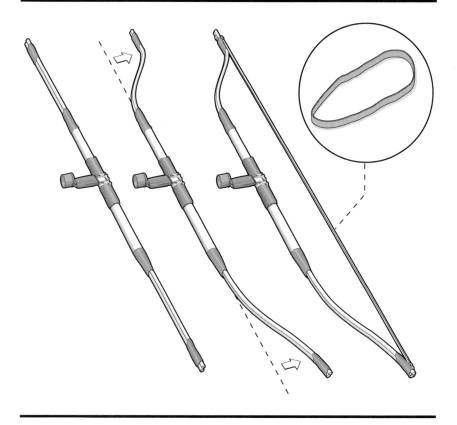

With the bow assembly complete, only the frame adjustment remains. With your hands, slowly bend the upper and lower limb of the metal bow backward, with a slight curve at the tip of the limbs. This bow shape–called recurve–is just a suggestion; the shape can be customized to the individual user.

For this bow, you will be using a nontraditional rubber band bowstring. Cut a large rubber band open, then tie one end around each tip of the bow–between the metal limbs. If a large rubber band is unavailable, two small rubber bands can be knotted together, or string can be substituted for realism.

Step 6

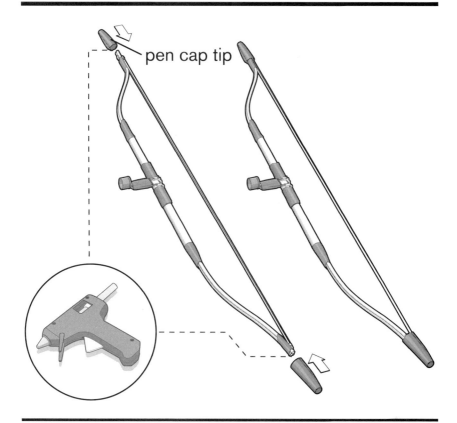

pen cap tip

To protect everyone from the cut metal coat hanger ends, carefully hot glue both round pen cap tops to the ends of the bow limbs. As a bonus, this detail will add a nice look to the bow design.

Step 7

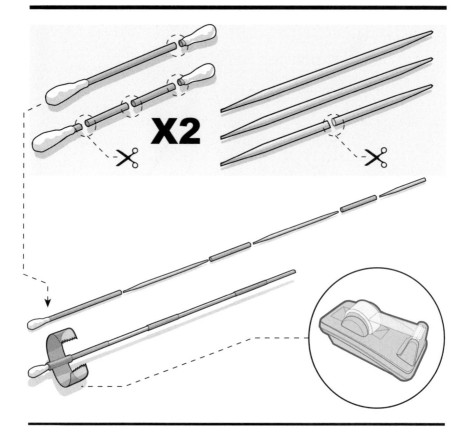

The soft-tipped arrows will be constructed from three toothpicks and the hollow shafts of two plastic cotton swabs. Cut off three of the four ends of the cotton swabs using scissors, then cut the cotton swab stick with the two removed ends in half.

Cut one of the three toothpicks in half. Slide each toothpick point into the plastic cavity of the cotton swab, staggering the toothpicks in this order: cotton swab with one cotton end, toothpick, half cotton swab stick, toothpick, half cotton swab stick, half toothpick. The half toothpick will be the nock (end).

To increase the arrow's accuracy, add weight to the point. Tightly wrap a 12-inch piece of clear tape around the toothpick about ½ inch from the point. You may need to decrease or increase the amount of tape until you arrive at the proper balance.

When you're ready to fire, use the pen clip arrow guide to increase accuracy. Slowly pull back the bowstring and arrow, aim, and release.

Consult the safety instructions on page ix!

Bows and Arrows

COMPOSITE RULER BOW

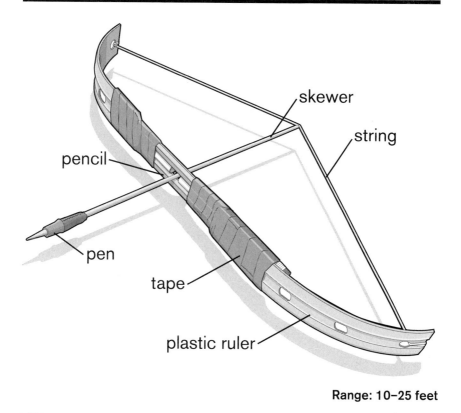

skewer

string

pencil

pen

tape

plastic ruler

Range: 10–25 feet

Powerful and durable, the Composite Ruler Bow is designed for the archer on the move! Its wooden pencil core gives it dimensional stability and durability, while the plastic limbs store the majority of the bowstring's energy. Joined together, these everyday materials work seamlessly toward the same goal: hitting the target!

Supplies

1 plastic ballpoint pen
2 wooden pencils
Duct tape
1 plastic ruler with peg holes
String

Tools

Large scissors or hobby knife

Ammo

1+ wooden skewers

Step 1

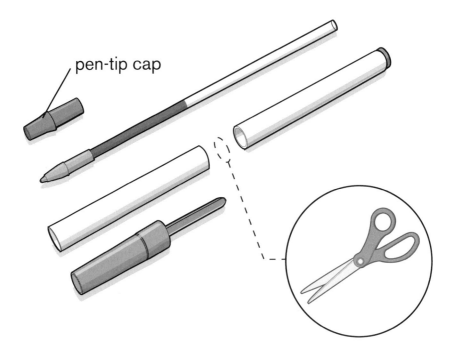

pen-tip cap

Disassemble one plastic ballpoint pen into its various parts. Separate the pen-tip cap from the ink cartridge. (The rear pen-housing cap does not have to be removed for this project.) Once the ink cartridge is removed, carefully cut the pen housing in half with large scissors or a hobby knife.

Step 2

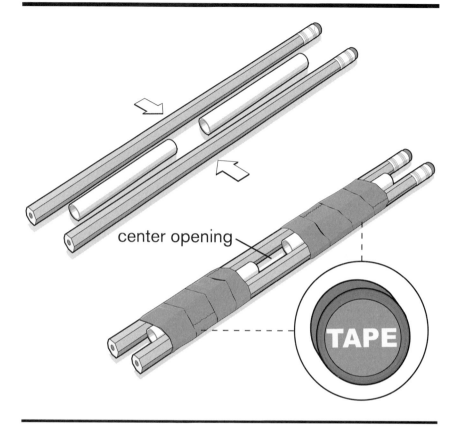

center opening

TAPE

Sandwich the pen halves between two pencils, leaving ½ inch between the pen halves. Fasten the assembly with tape as shown. Do not tape over the center opening.

Step 3

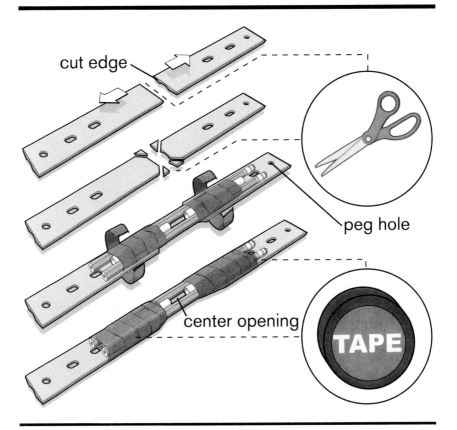

cut edge

peg hole

center opening

TAPE

With the large scissors or hobby knife, carefully cut a plastic ruler in half. On the cut edge, trim both corners on each ruler half at a small, ¼-inch angle. These cuts will eliminate sharp edges and help the bow design transition into the wooden core.

On opposite ends of the pencil assembly, and with the ruler's peg holes facing outward, overlap each section of ruler approximately 2 inches onto the wooden frame, then tightly tape all three segments together. Again, do not cover up the center opening of the frame.

Step 4

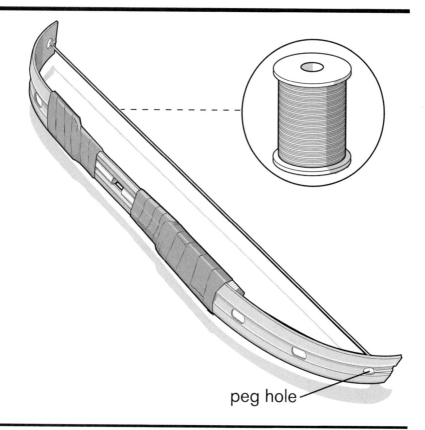

peg hole

Attach a string to one end of the bow, using the peg hole as the tie-down. Use a double knot. Then, with a slight bend in the bow, tie the string to the opposite peg hole, making sure the bowstring is tight. Trim any excess string at each end.

No string? No problem—substitute a rubber band for the same power!

Step 5

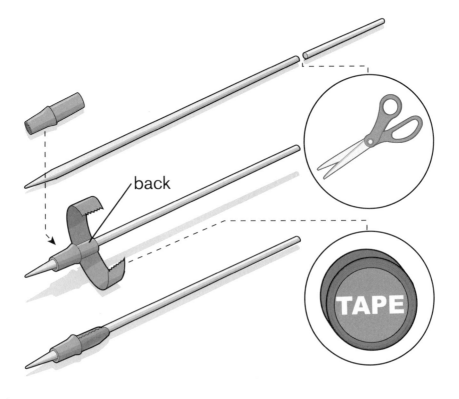

back

The shaft of the arrow will be constructed from a wooden skewer. Cut the skewer so that its length is roughly 8 inches. To increase accuracy, add weight to the skewer's tip. Slide the pen-tip cap left over from step 1 approximately ½ inch onto the tip of the skewer. Tape the back of the pen-tip cap in place. The arrow is complete. (If a wooden skewer is unavailable for this project, refer to the arrows used in the other bow and crossbow projects in this book.)

To fire, use the opening in the bow frame as a guide, then draw the bowstring, aim, and let it fly! *Remember to use eye protection!* Arrows can travel at a high velocity and have sharp points. *Mini-Weapon projects are not meant for living targets*. Always stay clear of spectators and operate the bow in a controlled manner. Home-made weaponry can malfunction.

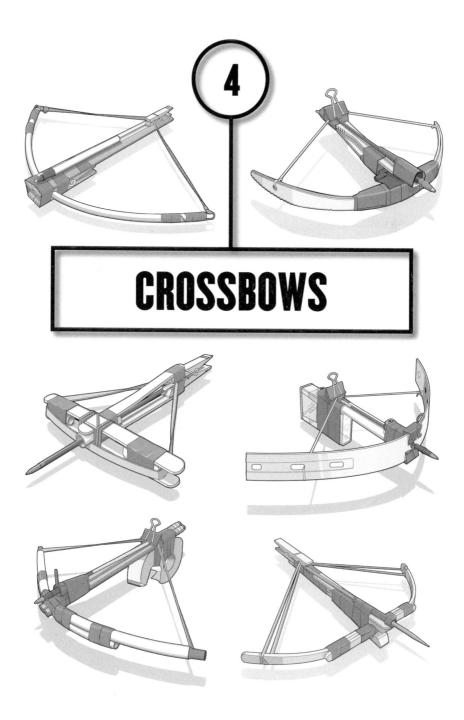

CROSSBOWS

WOODEN RULER CROSSBOW

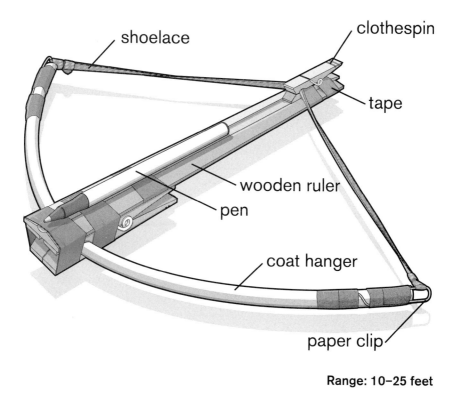

shoelace

clothespin

tape

wooden ruler

pen

coat hanger

paper clip

Range: 10–25 feet

Speak softly but carry a big crossbow! With its long wooden stock and powerful plastic bow, the Wooden Ruler Crossbow is just that. Its simple design requires minimal materials but offers maximum fun. With features like a built-in trigger and a durable frame, it's the perfect crossbow for beginners.

Supplies

1 plastic coat hanger
2 large paper clips
Duct tape
3 wooden clothespins
1 wooden ruler
1 shoelace

Tools

Safety glasses
Wire cutters
Hot glue gun (optional)
Scissors

Ammo

1+ plastic ballpoint pens or
    wooden pencils

Step 1

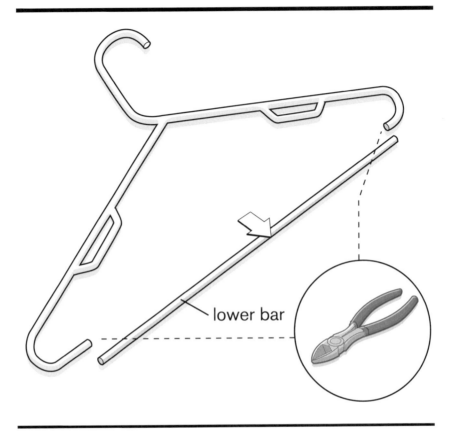

lower bar

Using wire cutters, remove the lower bar from a plastic coat hanger. This bar will be used as the bow. The remaining materials should be recycled.

Step 2

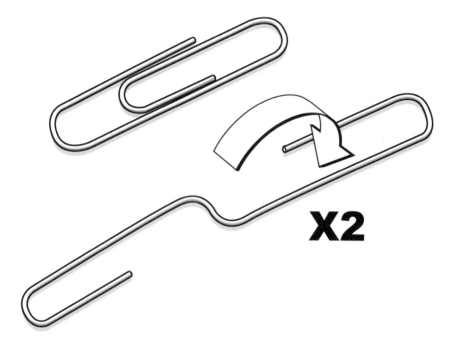

X2

Bend open two large paper clips 180 degrees to double their length, creating "S" shapes.

Step 3

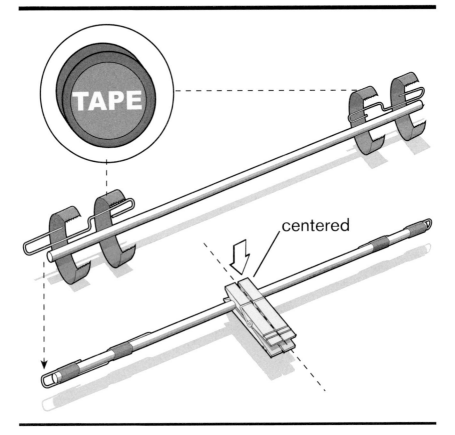

Tape the modified paper clips to the opposite ends of the plastic hanger's lower bar. Both clips should protrude ¼ inch from the opposite ends of the hanger rod.

Due to the stress these clips will endure, add more tape to the modified paper clips to prevent malfunction during use. The illustration shows where to place the tape.

Next, clamp two wooden clothespins at the center point of the bar.

Step 4

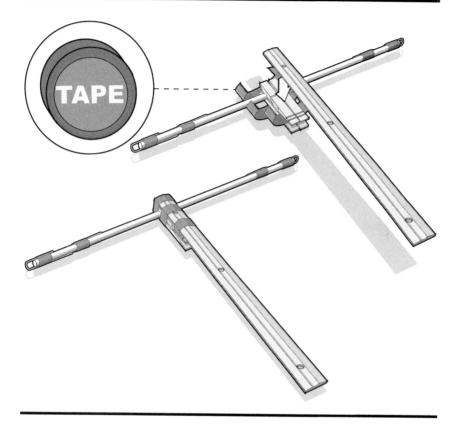

TAPE

Place a wooden ruler on top of the two attached clothespins. The end of the ruler should align with the front edge of the clothespins. Secure the ruler to the clothespins with multiple pieces of tape.

For a cleaner look, hot glue can be substituted for the tape.

Step 5

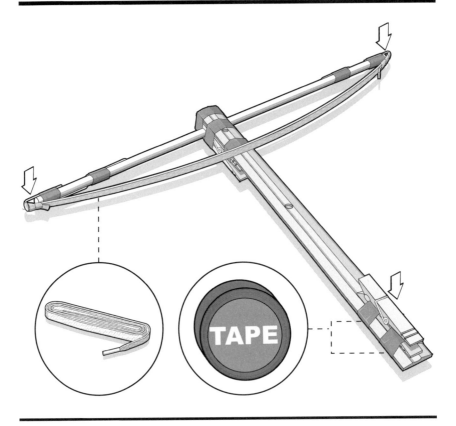

For the trigger, add the last clothespin on top of the back of the attached wooden ruler. Tape the clothespin by the *bottom prong only*. The clothespin must still function properly after attached.

To finish the Wooden Ruler Crossbow, tie one end of a shoelace (or string) to the paper clip loop attached to the bow assembly; a double knot works well. Next, slightly bend the bow to add tension and then tie the opposite end of the bow with another double knot. Use scissors to remove any excess shoelace from both sides of the bow.

To fire, pull the shoelace into the clothespin trigger, load a pen or pencil onto the ruler's flight groove, and fire at will. **Remember to use eye protection!** Bolts (arrows) can travel at a high velocity and have sharp points. ***MiniWeapon projects are not meant for living targets***. Always stay clear of spectators and operate the bow in a controlled manner. Homemade weaponry can malfunction.

PLASTICWARE CROSSBOW

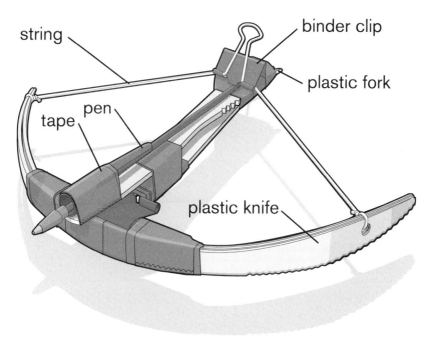

string

binder clip

plastic fork

tape pen

plastic knife

Range: 10–25 feet

Short and to the point, the Plasticware Crossbow requires few supplies and is inexpensive to manufacture. The short stock reduces the draw length, which is offset by a unique double bow that increases the energy released. Hold the drawn bowstring with the trigger until it's go time!

Supplies

1 plastic pen cap
1 plastic fork
Duct tape
5 plastic knives
1 medium binder clip (32 mm)
1 small binder clip (19 mm)
String

Tools

Safety glasses
Large scissors or hobby knife
Power drill (optional)

Ammo

1+ pen ink cartridges, wooden
    skewers, or thin dowels

Step 1

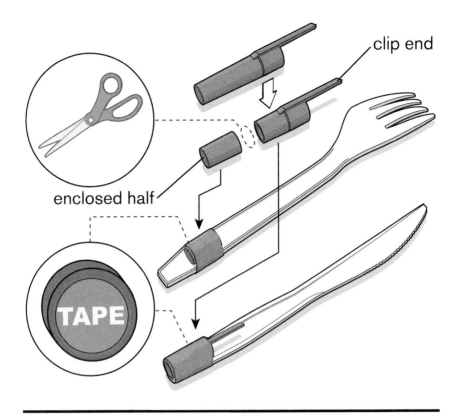

clip end

enclosed half

TAPE

Start construction of the Plasticware Crossbow by carefully cutting a plastic pen cap in half using large scissors or a hobby knife, as illustrated.

Tape the enclosed half of the pen cap to the underside of a plastic fork's handle. Position the cap approximately 1 inch from the end of the handle. The second half of the pen cap—the clip end—should be taped to the end of the handle of a plastic knife. Place the cap with the clip detail pointing toward the cutting edge and aligned as shown, and the cut end of the cap flush with the end of the handle.

Both of these assemblies will be used for the crossbow stock. Set them aside till step 4.

Step 2

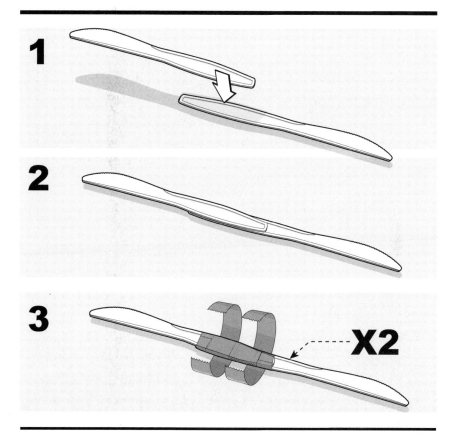

Four plastic knives will be used in this step. Overlap two knife handles as shown. Each handle should overlap approximately 3 inches (images 1 and 2). Once in place, tightly tape both knives together where the handles overlap (3). Then repeat the steps in images 1 through 3 for an additional knife assembly.

Step 3

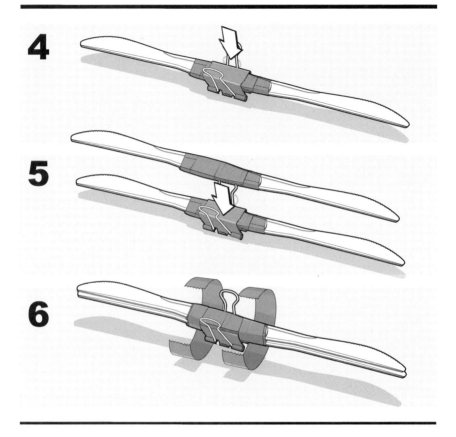

Clamp one medium binder clip to the center of one of the knife assemblies (4). Then place the second knife assembly onto the backside of the binder clip, opposite the clip's opening (5). Both knife assemblies should be aligned.

Tightly secure the assemblies by taping both sides of the binder clip (6).

Step 4

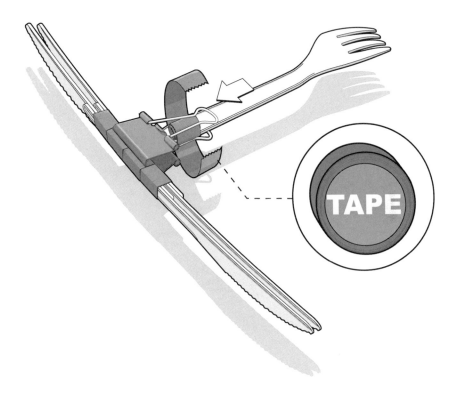

Slide the upside-down fork/pen cap assembly into the binder clip as shown. Fold down the metal binder clip handles so they rest on the top and bottom surface of the fork assembly. Once in place, tape around the fork and metal handles.

Step 5

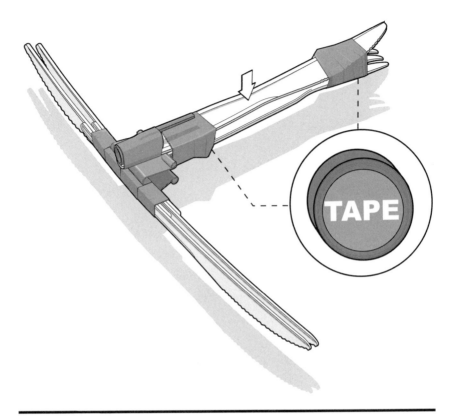

Place the knife assembly from step 1 on top of the upside-down fork. The attached pen cap detail should be pointing toward the bow end, with the knife overhanging the binder clip by approximately ¼ inch.

With tape, bind the knife assembly and frame together. Place the first piece of tape underneath the attached pen cap clip, but do not obstruct the opening of the pen cap. The second piece of tape should be placed around the cutting edge.

Step 6

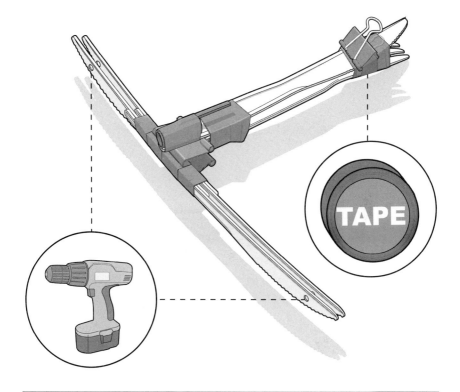

On opposite ends of the bow assembly, approximately ½ inch from the knife tips, drill two string-sized-diameter holes. The plastic housing can crack, so drill slowly and use a new drill bit. No drill? No problem—the string can still be attached to the limbs in the next step using only tape (not shown).

Next, tape one small binder clip to the back of the plasticware stock, above the knife cutting edge. Secure the metal clip in place by taping around the bottom metal handle first, and then add more tape through the binder clip frame. The binder clip must still function properly after attached.

Step 7

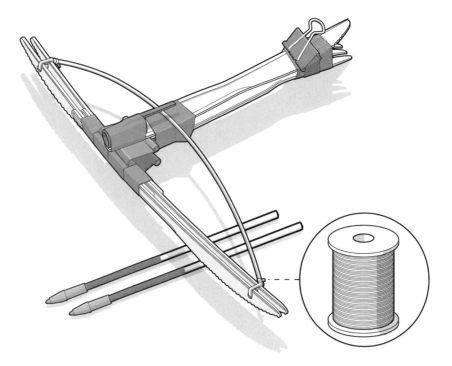

To finish the Plasticware Crossbow, you need to add the bowstring. Tie string to one end of the bow utilizing both drilled holes from the previous step; a double knot is recommended. Next, with the bow slightly bent, tie the opposite end of the bow with another double knot through the opposite hole, keeping the string tight. Trim any excess string from both sides. The Plasticware Crossbow is complete!

To fire, pull back the string into the binder clip and then load an ink cartridge, wooden skewer, or thin dowel, using the pen cap as a guide. Carefully pick a target, aim, and release the binder clip trigger.

Remember to use eye protection! Bolts (arrows) can travel at a high velocity and have sharp points, and plastic utensils will break under severe stress. ***MiniWeapon projects are not meant for living targets***. Always stay clear of spectators and operate the bow in a controlled manner. Homemade weaponry can malfunction and ***ink cartridges can explode on impact***.

CRAFT STICK CROSSBOW

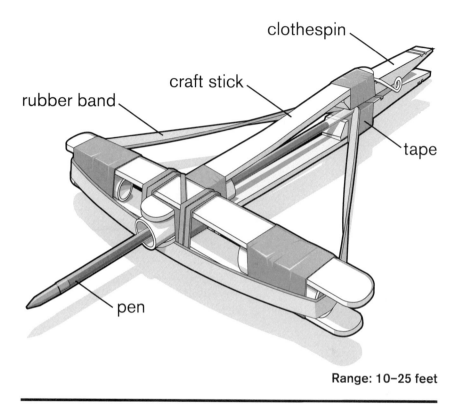

clothespin

craft stick

rubber band

tape

pen

Range: 10–25 feet

Hailed as a superior weapon, the medieval crossbow did have some disadvantages. Outfitting an entire army with crossbows was very expensive, even for the wealthiest of kings, and manufacturing the weapon took critical time. Here's a modern-day solution: the Craft Stick Crossbow! Inexpensive to mass produce, it's a great addition to any auxiliary regiment.

Supplies

1 plastic ballpoint pen
4 craft sticks
Duct tape
1 wooden clothespin
1 wide rubber band
1 rubber band

Tools

Large scissors or hobby knife
Small pliers (optional)

Ammo

1+ pen ink cartridges

Step 1

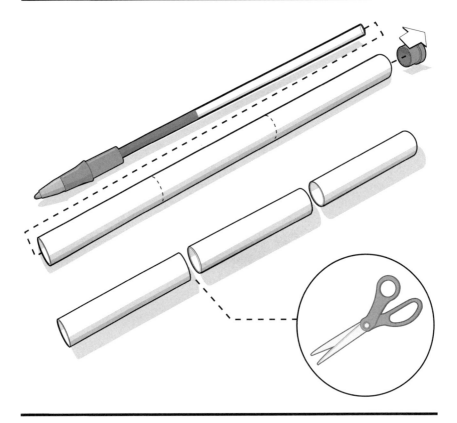

Disassemble a plastic ballpoint pen into its various parts. Depending on how the pen has been manufactured, you may need a tool to help dislodge the rear pen-housing cap. Large scissors or a hobby knife (for cutting it off) or small pliers (for pulling it out) both work.

With the ink cartridge removed, use the large scissors or hobby knife to carefully cut the pen housing into three equal parts.

Step 2

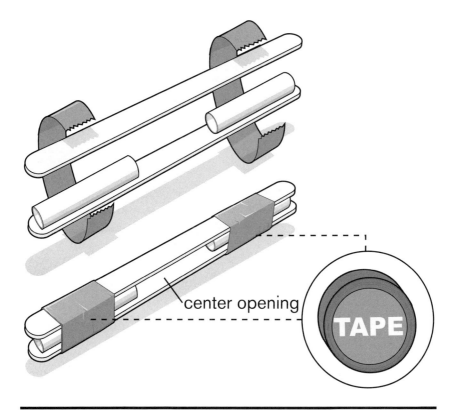

center opening

TAPE

At a distance of ¼ inch from both ends, sandwich the pen housing parts between two craft sticks as shown. Tape the assembly together, but do not tape over the center opening. The nonflexing crossbow limb is complete.

Step 3

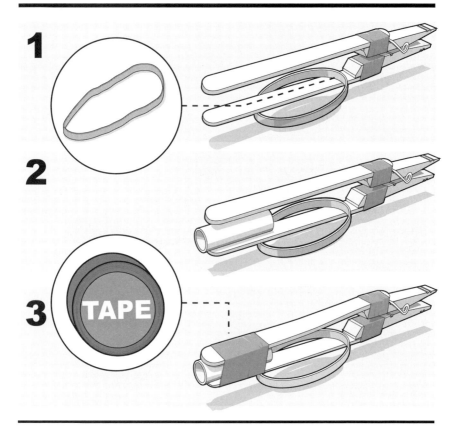

Next, with a 1-inch overlap, tape two craft sticks onto the top and bottom clothespin prongs. *Do not tape around the entire clothespin when assembling*–the clothespin should still work after assembly. Slide a wide rubber band between the two assembled craft sticks (1). This rubber band will remain loose for now.

Place the third and last pen housing part between the two attached craft sticks (2). With the cylinder flush to the tips of the craft sticks, tightly tape the assembly together (3). The crossbow frame is complete.

Step 4

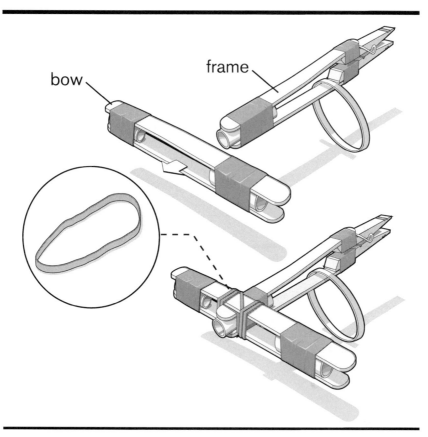

It's time to combine the frame and bow assemblies from the previous steps. First, slide the frame assembly between the craft sticks in the bow housing. With the frame centered and pen housing extending about ¼ inch from the bow, attach both assemblies with one rubber band, as shown.

Step 5

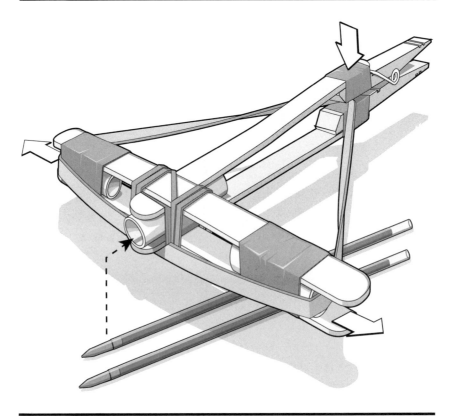

Now loop the dangling rubber band around the bow assembly as shown. To prevent obstruction, tuck the front loop of the rubber band under the pen housing's ¼-inch overhang. The rear loop functions as a nontraditional bowstring. The Craft Stick Crossbow is complete.

To fire, draw (pull) the rubber band into the clothespin, locking the bowstring into place. Then load a pen ink cartridge into the pen housing, as shown on page 157, carefully pick a target, and release the rubber band.

Remember to use eye protection! Bolts (arrows) can travel at a high velocity and have sharp points. Rubber bands will break under severe stress. ***MiniWeapon projects are not meant for living targets***. Always stay clear of spectators and operate the bow in a controlled manner. Homemade weaponry can malfunction and ***ink cartridges can explode on impact***.

PLASTIC RULER CROSSBOW

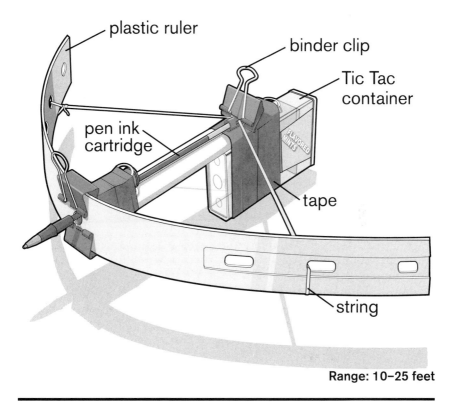

plastic ruler

binder clip

Tic Tac container

pen ink cartridge

tape

string

Range: 10–25 feet

Intimidate the opposition with the amazing Plastic Ruler Crossbow. With its durable plastic frame and metal clip fittings, this crossbow is built to last. At the end of the stock is an integrated handle that increases control, which helps keep the archer's aim true.

Supplies

2 plastic ballpoint pens
1 plastic ruler with peg holes
3 small binder clips (19 mm)
Duct tape
1 Tic Tac container
String

Tools

Safety glasses
Pliers or a thin dowel (optional)
Hobby knife
Scissors (optional)

Ammo

1+ pen ink cartridges

Step 1

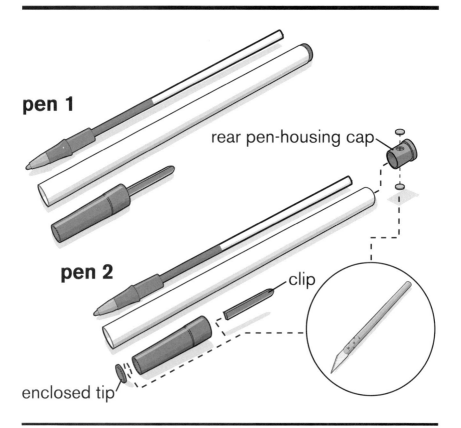

pen 1

rear pen-housing cap

pen 2

clip

enclosed tip

Disassemble two plastic ballpoint pens by removing the tips and ink cartridges.

Dislodge the rear pen-housing cap from one of the pens. You may need a tool—pliers or a thin dowel—to dislodge the cap. Modify the removed rear pen-housing cap with a hobby knife by cutting two small string-sized-diameter holes into the cap, parallel to one another.

Next, with the hobby knife, remove both the clip and the enclosed tip from the pen cap.

Step 2

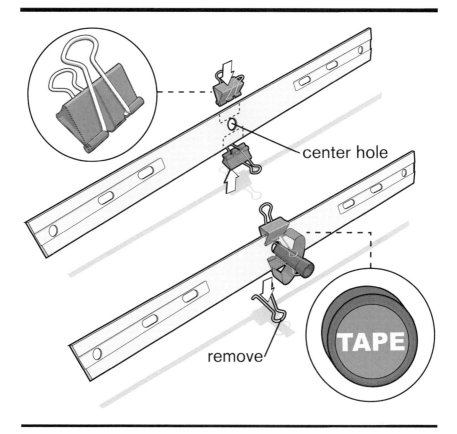

center hole

TAPE

remove

When picking a plastic ruler for this project, choose one that has several peg holes molded into the ruler housing, including two at opposite ends of the ruler and one in the center. Clamp two small binder clips to the ruler's center, one above the peg hole and the other below it. The clip housings should not block the hole.

Then slide the trimmed pen cap between the two clips as illustrated, resting the cap between two inner metal handles attached to the binder clips. Tape the pen cap/metal handle assembly together. Of the two remaining metal handles, keep the top one in place and remove the bottom one, as shown.

Step 3

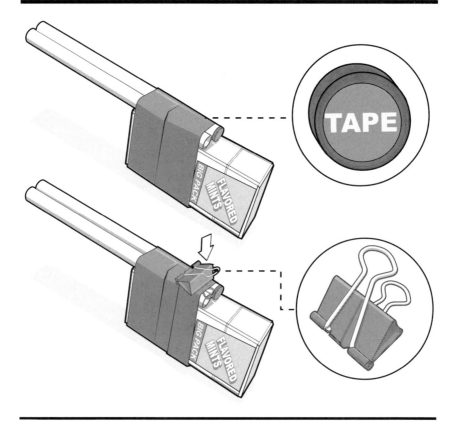

The crossbow handle will be constructed from a Tic Tac container. Place both pen housings on one side of the Tic Tac container, with an overlap of approximately 1½ inches. Tape the pen housings onto the container.

Next, tape one small binder clip to the back of the stock, above the pen housing.

Secure the clip in place by taping around the metal handle, and then through the inside of the binder clip frame. The clip must still function properly after attached.

Step 4

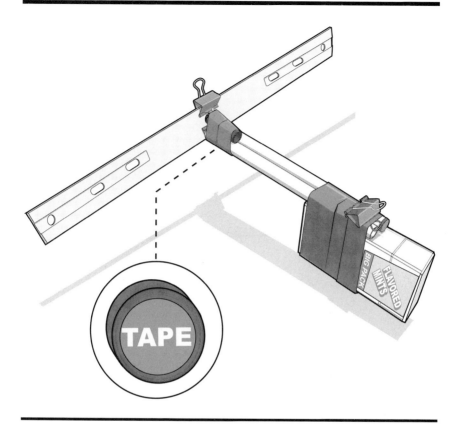

Now place the ruler assembly onto the front of the frame by resting the attached pen cap onto the ends of the pen housings. Straighten the ruler assembly to a 90-degree angle, then tightly wrap tape around the cap and pen housings. This connection will malfunction if not securely taped, so double-check the connection before continuing.

Step 5

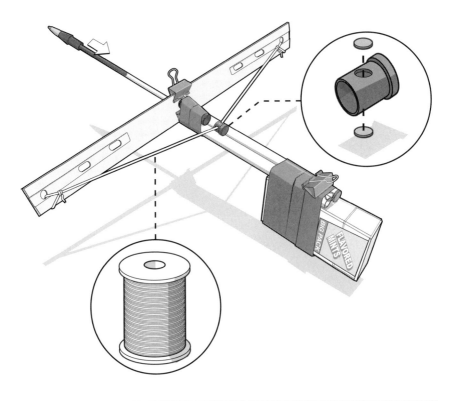

To finish the crossbow, the bowstring must be added. Start by double-knotting string through the peg hole at one end of the ruler. A double knot is suggested for the anticipated tension. Then feed the attached string through both holes in the rear pen-housing cap that you cut in step 1. This pen-housing cap will help with the firing mechanism.

Next, with a slight bend in the ruler, tie the opposite end of the bow, making another double knot through the other peg hole and keeping the string tight.

The Plastic Ruler Bow is complete. To fire, load a bolt (ink cartridge) into the pen-housing cap and then lock the cap into the binder clip. Carefully pick a target, aim, and then release the binder clip trigger.

Consult the safety instructions on page ix! Bolts (arrows) can travel at a high velocity and have sharp points, a plastic ruler can break under severe stress, and ***ink cartridges can explode on impact***.

OFFICE CROSSBOW

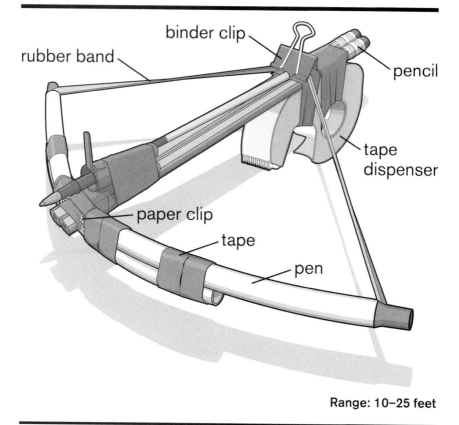

binder clip

rubber band

pencil

tape dispenser

paper clip

tape

pen

Range: 10–25 feet

It's every MiniWeapon user's right to protect his or her desktop domain from invaders. Fight off any stapler-stealing thief with the Office Crossbow! Built with everyday items—pens, pencils, paper clips, a binder clip, and a tape dispenser—it will be one of your proudest office accomplishments. And with its simple construction, there's no need to put in overtime.

Supplies

3 plastic ballpoint pens
4 large paper clips
2 wooden pencils
Duct tape
1 tape dispenser
1 wide rubber band
1 small binder clip (19 mm)

Tools

Safety glasses
Large scissors
Pliers
Hot glue gun (optional)

Ammo

1+ pen ink cartridges

Step 1

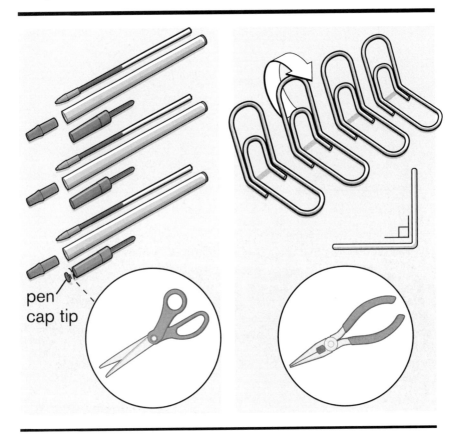

pen cap tip

Disassemble three plastic ballpoint pens by removing the tips and ink cartridges. Modify one of the pen caps by carefully cutting off the pen cap tip with large scissors as shown.

Next, with pliers, bend four large paper clips in half, creating one 90-degree angle in the center of each of them.

Step 2

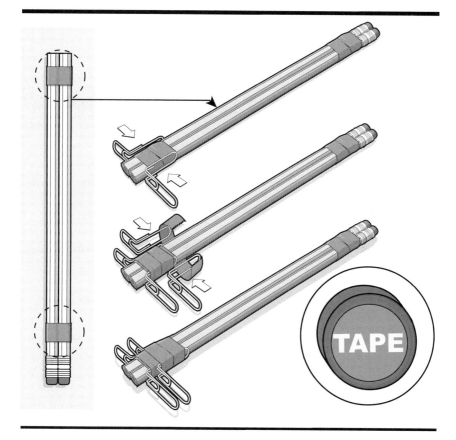

Begin the crossbow frame construction by taping two pencils together approximately ½ inch from each end of the grouping (left illustration).

Now tape two of the bent paper clips on opposite side of the pencil assembly approximately ¼ inch from the non-eraser end. Repeat this step by placing the second set of paper clips so the protruding segments are ¼ inch behind the previously attached set—approximately the width of the pen housings that will be attached in the next step.

If these connections are not securely taped, a malfunction will occur, so double-check the connections before continuing.

Step 3

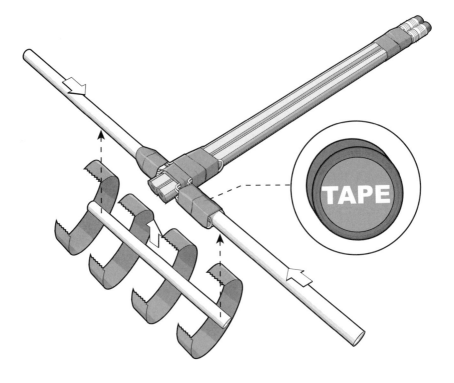

TAPE

On opposite ends of the pencil frame, slide two pen housings in between the attached paper clips. Tightly tape both pen housings to their respective paper clip supports.

To further support the bow assembly, tightly tape the third pen housing to the underside of the attached pen housings. The third pen housing should be centered prior to taping.

Step 4

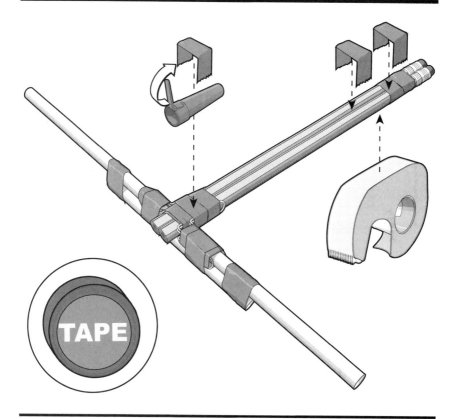

TAPE

On the modified pen cap (missing tip), slowly bend back the plastic clip detail to a 90-degree angle. Then tape this pen cap to the front of the crossbow frame with the clip detail pointing straight up so it can be used as a firing sight.

For the crossbow stock, tape an upside-down plastic tape dispenser to the rear of the frame. This dispenser will be a quick-and-easy handle, complete with a makeshift trigger guard.

Step 5

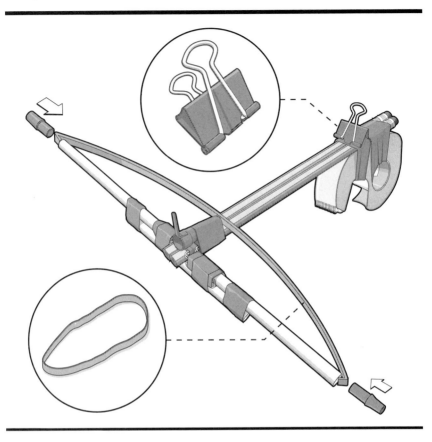

With scissors, cut open a wide rubber band. Then slide each end of the straightened rubber band into the hollow pen housings located at the opposite ends of the bow. To hold this bowstring in place, wedge both pen-tip caps back into the pen housings. If needed, add knots, tape, or glue to increase the hold.

Next, tape one small binder clip to the back of the stock, above the tape dispenser. Secure the clip in place by taping around one of the metal handles, and then through the inside of the binder clip frame. The clip must still function properly after attached.

The Office Crossbow is complete! To fire, load a bolt (ink cartridge) through the pen cap barrel and then lock the rubber band and ink cartridge into the binder clip. Carefully pick a target, aim and then release the binder clip trigger. ***Consult the safety instructions on page ix!*** Homemade weaponry can malfunction, and ***ink cartridges can explode on impact***.

COMPOUND CROSSBOW

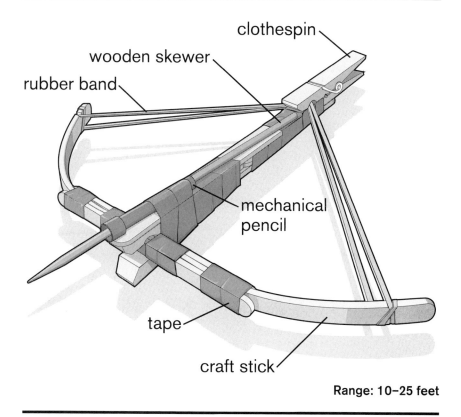

clothespin

wooden skewer

rubber band

mechanical pencil

tape

craft stick

Range: 10–25 feet

The traditional all-wooden design of the Compound Crossbow keeps speed, performance, and comfort in mind. Its custom bent limbs, which assist the "latex-powered" high-energy bowstring, make this Mini-Weapon the preferred crossbow for any highly skilled target shooter.

Supplies

7 craft sticks
5 rubber bands
2 wooden clothespins
1 inexpensive mechanical pencil
Duct tape

Hobby knife
2 2-peg-by-4-peg building blocks
Bowl of warm water
Large scissors (optional)

Ammo

1+ wooden skewers

Tools

Safety glasses

Step 1

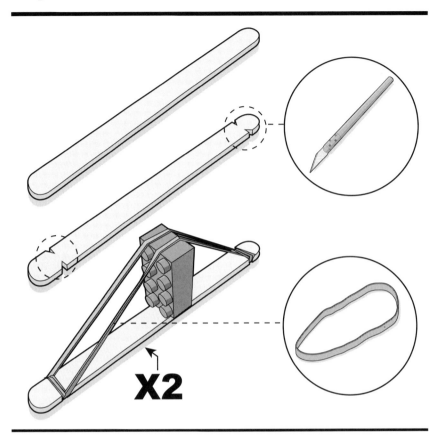

With a hobby knife, carefully cut two small wedge details out of each end of a craft stick, approximately ½ inch from the tips. Then loop one rubber band around both groups of notches. Once the band is in place, slide a 2-peg-by-4-peg building block (or something equivalent in size) upright, under the rubber band. Repeat this step with another craft stick, for a total of two finished assemblies.

Step 2

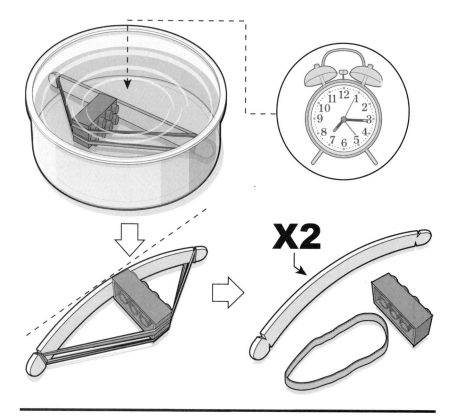

Wood bending with water is a fascinating process used by all types of woodworkers. You will sample this craft by fully submerging both craft stick assemblies into a full bowl of warm water, saturating the wood and giving it additional pliability. Soak the assemblies for 60 to 90 minutes before removing and bending them. The rubber band should have assisted with the bending process.

With the rubber band and block still attached, slowly increase each stick's arc with your fingers. Let the wet craft sticks dry for 30 minutes or more to allow the arc to set. Once the sticks are dry, remove the rubber bands and blocks.

Step 3

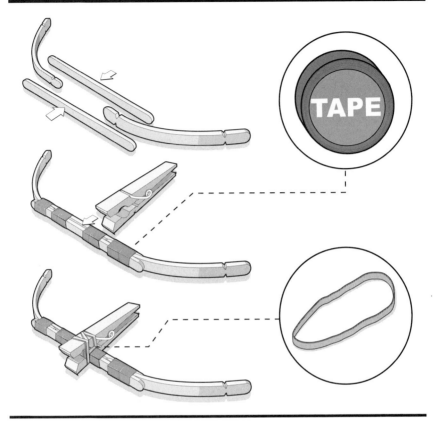

Sandwich the curved craft sticks between two straight craft sticks, at opposite ends approximately 2 inches apart, as shown in the top illustration. Tape the assembly together.

Next, clamp one clothespin to the center of the craft stick assembly. Wedge the craft stick assembly into the round detail of the clothespin prongs. Rubber band the clothespin and craft stick assembly together, as indicated.

Step 4

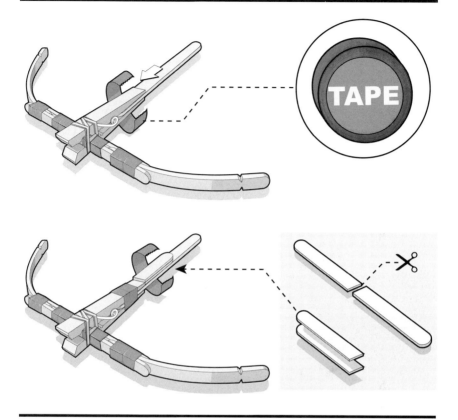

Slide one craft stick between the rear prongs of the attached clothespin. Once it is aligned with the frame and touching the metal spring, securely tape the craft stick into place.

With the hobby knife or a pair of large scissors, carefully cut another craft stick in half. Stack both halves behind the attached clothespin prong and rear craft stick. Tape them into place.

Step 5

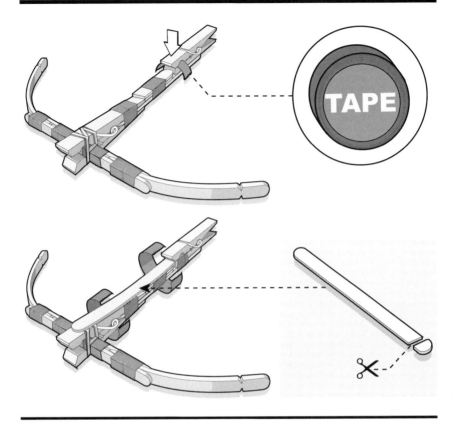

Next, tape the last clothespin to the back of the frame, resting it on the top rear of the attached craft stick. Secure the clothespin by wrapping tape through the inside of the clothespin's bottom prong. The clothespin must still function properly after attached.

Use the hobby knife or large scissors to carefully cut off ¼ inch of the end of the last remaining craft stick to remove its round tip. Then, with the blunt end against the rear clothespin, fasten the cut craft stick to the frame by taping it in multiple areas.

Step 6

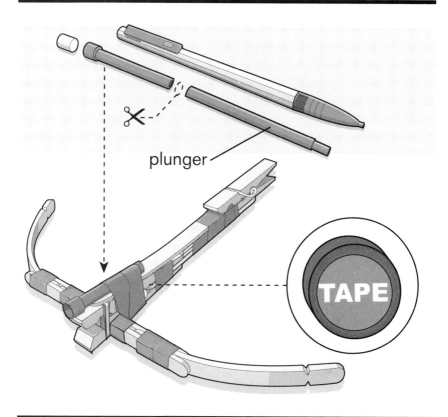

plunger

TAPE

Next, disassemble an inexpensive mechanical pencil using brute strength. Pull out the plunger and remove the attached eraser.

With the hobby knife or large scissors, carefully remove 1½ inches from the back of the plunger (eraser side). Place the 1½-inch plunger piece on top of the front clothespin, as illustrated, with the eraser housing facing forward. Tape the plunger into place.

Step 7

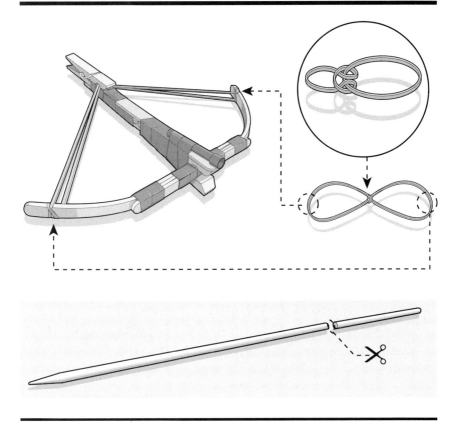

It's time to add some elastic firepower to the Compound Crossbow! If you have average-sized rubber bands, loop and knot two of them together; a single large-diameter rubber band can be used instead.

To start, double loop each side of the rubber band assembly to the notch detail at the ends of the bow limbs. The rubber band should stay in place during operation, but if the rubber band connection fails, use two additional rubber bands to secure the mounts.

The recommended bolt (arrow) for this crossbow is a shortened wooden skewer, cut down to 6 inches, but you can also use one of the other bolts or arrows suggested elsewhere in this book.

Construction is complete! To fire, lock the elastic bowstring into the clothespin and then load the skewer bolt through the mechanical pen barrel, resting the blunt end in front of the clothespin trigger. Carefully aim and release. **Consult the safety instructions on page ix!** Homemade weaponry can malfunction.

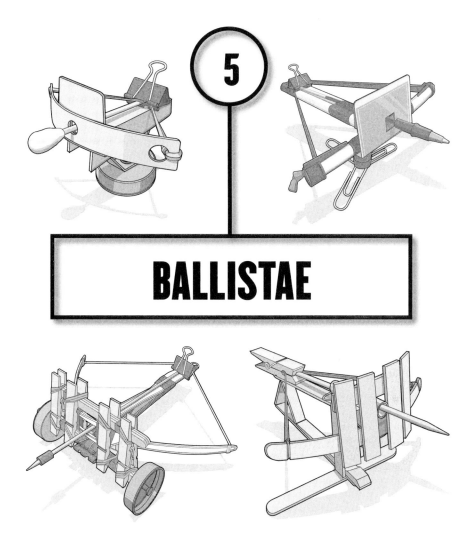

BALLISTAE

GIFT CARD BALLISTA

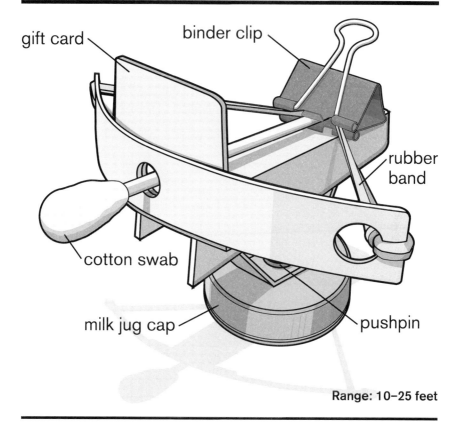

gift card

binder clip

rubber band

cotton swab

milk jug cap

pushpin

Range: 10–25 feet

Mounted on a rotating base, the Gift Card Ballista will prove its worth in any field commander's artillery collection. Constructed from an expired or zero-balance gift card, a pushpin, and a bottle cap, this ballista can be built cheaply by a kingdom lacking supplies and time.

Supplies

1 expired or zero-balance plastic gift card
1 small binder clip (19 mm)
1 plastic milk jug cap (or similar)
1 pushpin
1 wide rubber band

Tools

Marker
Scissors
Single-hole punch
Hot glue gun
Hobby knife (optional)

Ammo

1+ plastic cotton swabs

Step 1

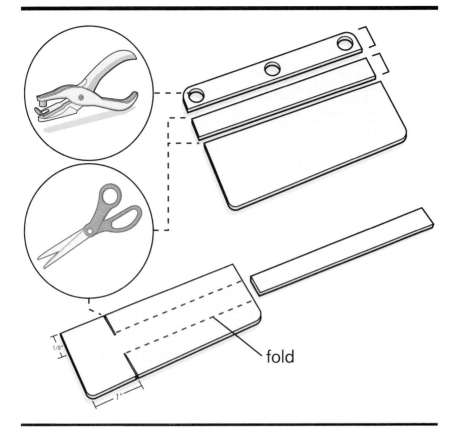

fold

Rummage through a junk drawer or junk mail to locate an expired or zero-balance plastic gift card. *The card will be destroyed, so do not use a card that is still active.*

With scissors, cut two ½-inch horizontal strips from the plastic gift card. Use a single-hole punch to make three holes in one of the plastic strips—two holes ⅛ inch from opposite ends and one hole dead center—as shown.

Center the other strip on top of the larger card section as a guide. Once in place, cut two 90-degree, ¼-inch slits opposite from one another approximately 1 inch from the end. The dotted lines represent the fold lines in step 2.

Step 2

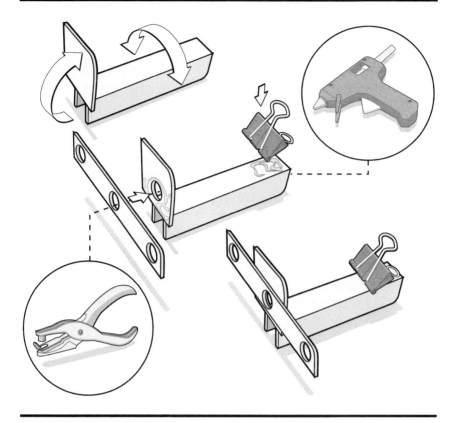

Using the ¼-inch slits, make three folds into the large, leftover card section. The front of the card will be folded up 90 degrees; the two slits are the fold lines for this bend. The second set of folds will be the back section; they should be folded down to a crisp 90-degree angle, parallel to one another.

With the hole punch, make one hole centered and approximately $5/16$ of an inch from the bottom of the front folded-up face. Then carefully hot glue the three-holed strip to the upright face, lining up the center holes.

Next, carefully hot glue the small binder clip to the back of the assembly as shown.

Step 3

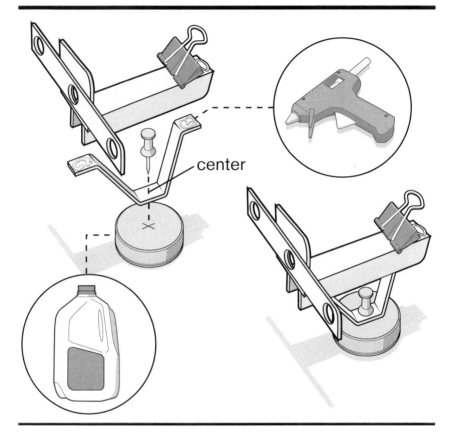

center

The other ¼-inch card strip will be bent four times into a functional support for the ballista. The first two 45-degree bends should be made approximately ¼ inch from opposite ends of the gift card strip. The second group of bends will be ¼ inch from the center of the card and at 45 degrees–opposite from one another. Once completed, the center of the card should be a ½-inch-wide area for the pushpin connection, as shown in the illustrations.

Connect the frame support (folded card) to the base (milk jug cap) by placing a pushpin through the center of the frame support into the milk jug cap. Once connected, carefully hot glue the ballista frame onto the folded support.

Step 4

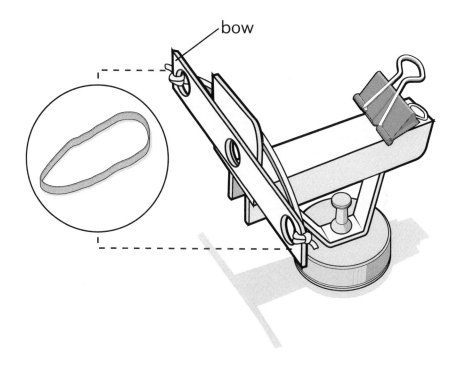

bow

For this ballista, you will be using a nontraditional rubber band bow-string. Cut open a wide rubber band, and then tie both ends of the rubber band to opposite ends of the bow, using the outer holes as the rubber band tie-downs. The mounted rubber band should have minimal slack.

Step 5

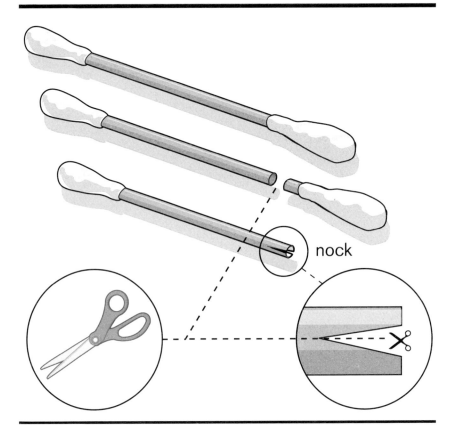

nock

This ballista will be armed with a less devious bolt than its real-life inspiration: a soft-tipped plastic cotton swab. Remove one end of a cotton swab with scissors. Then, with scissors or hobby knife, make one small slit in the cut end of the cotton swab handle, for the nock detail.

Time to attack! Load the modified cotton swab's cut end through the front firing hole, then slide the rubber band into the cotton swab nock and clamp both the rubber band and cotton swab into the binder clip. Wait for your target, and release.

Remember to use eye protection! Bolts (arrows) can travel at a high velocity and have sharp points—cotton swabs included. **Mini-Weapon projects are not meant for living targets**. Always stay clear of spectators and operate the bow in a controlled manner. Homemade weaponry can malfunction.

Ballistae

PEN BALLISTA

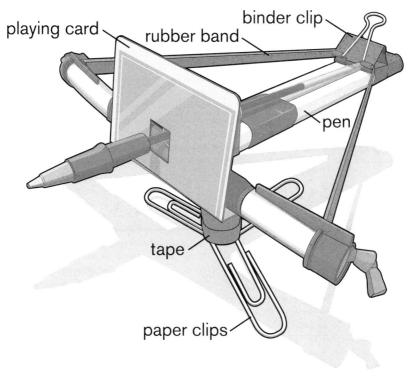

playing card · rubber band · binder clip · pen · tape · paper clips

Range: 10–25 feet

The Pen Ballista will quickly gain a reputation in the office or home as a formidable siege weapon. Built with the spoils from an office supply raid, it's a great way to recycle, reuse, and attack any target. Mounted on a tripod, this ballista can relentlessly launch ink cartridge bolts from a stationary base, increasing accuracy.

Supplies

3 plastic ballpoint pens
6 large paper clips
Duct tape
1 small binder clip (19 mm)
1 playing card
1 rubber band

Tools

Safety glasses
Large scissors or hobby knife
Wire cutters or pliers

Ammo

1+ pen ink cartridges

Step 1

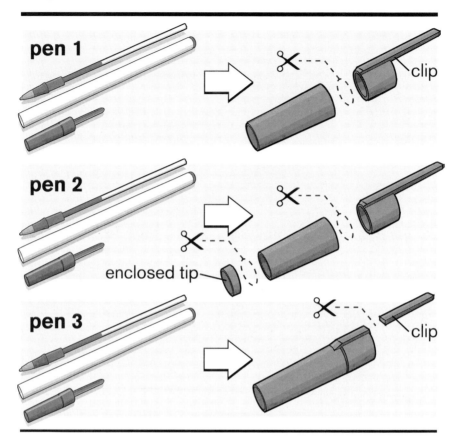

pen 1 — clip

pen 2 — enclosed tip

pen 3 — clip

Remove the pen cap, ink cartridge, and pen-tip cap from three plastic ballpoint pens. With large scissors or a hobby knife, carefully modify each pen cap as follows:

Pen 1: Cut the last ¼ inch off the pen cap, keeping the clip attached to the removed segment. The removed housing will be used in step 7.

Pen 2: Cut the last ¼ inch off the pen cap, keeping the clip attached to the removed segment. Then cut off the enclosed tip. The center segment will be used in step 5, and the ¼-inch clip detail will be used in step 7.

Pen 3: Cut only the clip detail off the pen cap. This pen cap will be used in step 8.

Step 2

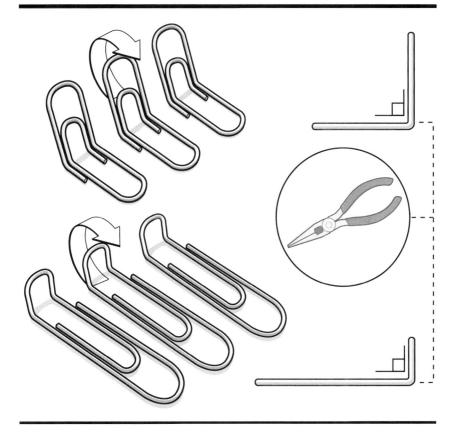

With wire cutters or pliers, bend three large paper clips, making one 90-degree angle at the center of each.

Then bend three additional large paper clips, each with one 90-degree angle approximately ¼ to ½ inch from one end of the clip.

Step 3

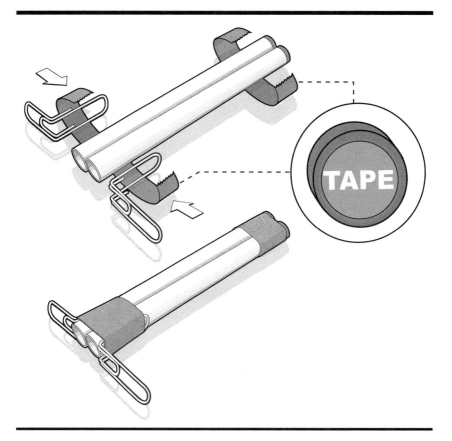

Start constructing the frame by taping two pen housings together. The pen housing ends should be neatly aligned with one another prior to taping.

Next, on opposite sides of the housings (but at the same end), tape two of the half-bent, 90-degree-angled paper clips flush to the end of the pens. Secure the paper clips (bow supports) by tightly wrapping tape around them.

Step 4

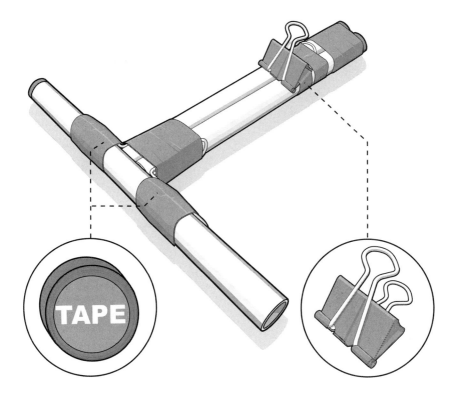

TAPE

Center and tape the third pen housing to the front of the attached paper clips.

Next, tape one small binder clip to the back of the pen frame. Secure the clip in place by taping around the bottom metal handle, and then through the inside of the binder clip frame. The clip must still open properly after attached.

Step 5

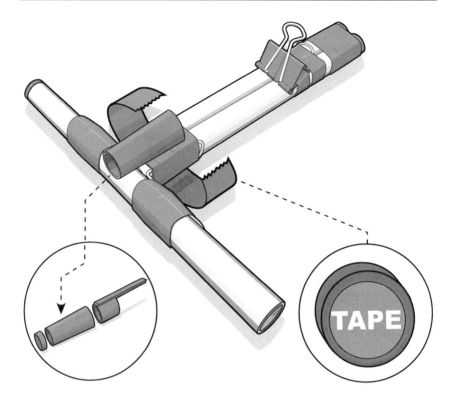

You will make a bolt guide (barrel) out of the center segment of the cap from pen 2. Place the hollow cap on top of the cross-section of the pen frame as shown, with the front of the pen cap flush with the pen housing bow. Secure the cap by tightly wrapping tape around the cap and frame assembly.

Step 6

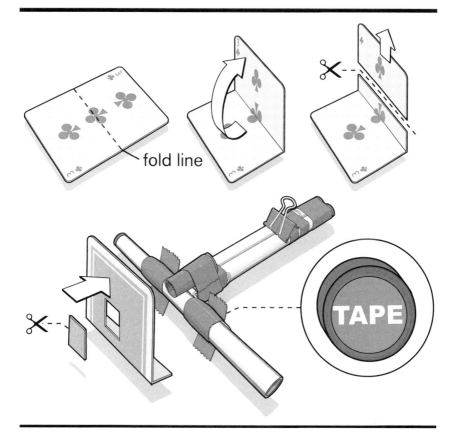

fold line

TAPE

Because mounted ballistae are mainly stationary weapons, a makeshift protective shield is sometimes added to the front of the siege weapon. This scale model will have a shield fabricated from a playing card.

Start by folding a playing card in half, making a 90-degree angle. Then cut the card ¼ inch from the 90-degree fold line, as shown.

For the last step, retrofit the card to the front of the pen. Use the scissors or hobby knife to cut out a ¼-inch box—about the size of the pen cap diameter—from the face of the card, then tape the finished playing card to the front of the pen assembly as shown.

Step 7

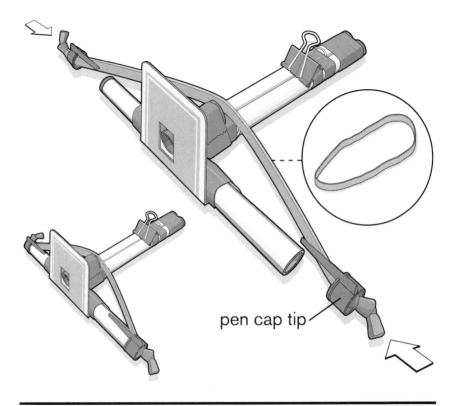

pen cap tip

This ballista will be powered by a rubber band, unlike the torsion bal-lista of medieval times. Cut open a rubber band, then tie two knots at opposite ends of the band. These knots will hold the bowstring in place and prevent malfunction during operation.

Attach the knotted rubber band onto the pen barrel by first sliding the rubber band ends through the two pen cap tips from step 1. Then slide each pen cap tip onto the opposite ends of the pen barrel, lock-ing the rubber band in place.

The ballista is finished. Next is the tripod assembly.

Step 8

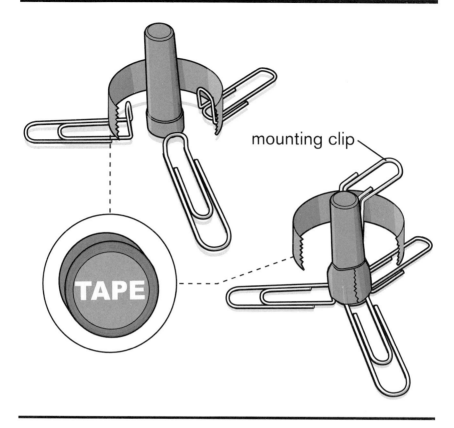

mounting clip

TAPE

Evenly space the three ¼-inch bent paper clips around one end of the pen cap from step 1, and then attach only the ¼-inch bend to the pen cap with tape. The attached pen caps will function as feet of a tripod; adjust them if needed.

Next, tape the last 90-degree-angled paper clip to the opposite end of the three-clip tripod. The 90-degree bend should be flush with the top of the pen cap.

Step 9

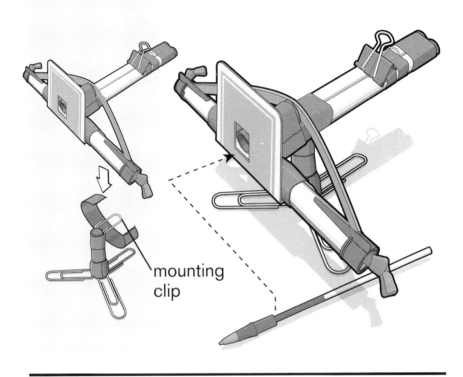

mounting clip

Finally, attach the tripod assembly to the underside of the ballista frame. To do this, line up the tripod so that the mounting clip is directly behind the attached pen cap barrel, then tightly tape the mounting clip to the pen frame. Once attached, adjustments to the mounting clip can be made to balance the attached ballista.

The ballista fires just like a crossbow. Lock the rubber band (bowstring) into the binder clip, load the ink cartridge bolt, and fire when ready. ***Remember to use eye protection!*** Bolts (arrows) can travel at a high velocity and have sharp points. ***MiniWeapon projects are not meant for living targets***. Always stay clear of spectators and operate the bow in a controlled manner. Homemade weaponry can malfunction and ***ink cartridges can explode on impact***.

CLOTHESPIN BALLISTA

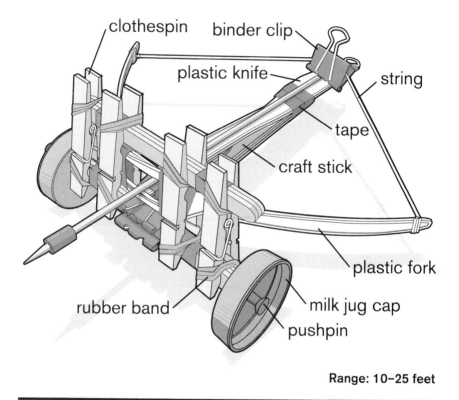

clothespin binder clip

plastic knife

string

tape

craft stick

plastic fork

rubber band

milk jug cap

pushpin

Range: 10–25 feet

This is it: the highly prized rolling Clothespin Ballista. Requiring no glue, this miniature, fully functional replica can be made for pennies!

Supplies

2 plastic forks
String
Duct tape
10 craft sticks
1 medium binder clip (32 mm or
    25 mm)
1 plastic knife
1 small binder clip (19 mm)
4 wooden clothespins
8 rubber bands
2 pushpins
2 plastic milk jug caps (or similar)

Tools

Safety glasses
Pliers (optional)
Large scissors
Power drill (optional)

Ammo

1+ wooden skewers
Clear tape

Step 1

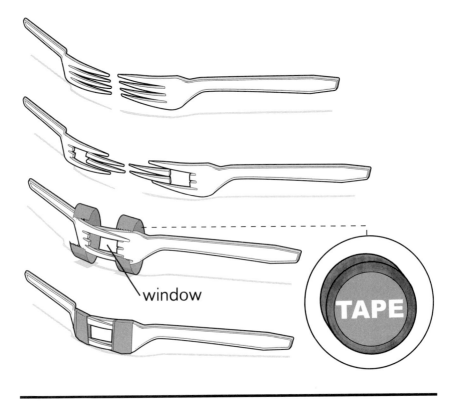

window

TAPE

Start with two plastic forks. With your fingers, snap off the two center prongs from both forks. Depending on the plastic's thickness, you may need a tool to remove the prongs; small pliers should be sufficient. Discard the removed prongs.

With the fork prongs pointing at one another, overlap the attached prongs to create a small window. Attach both forks together by tightly wrapping tape around the fork housing. Do not cover the opening with tape.

Step 2

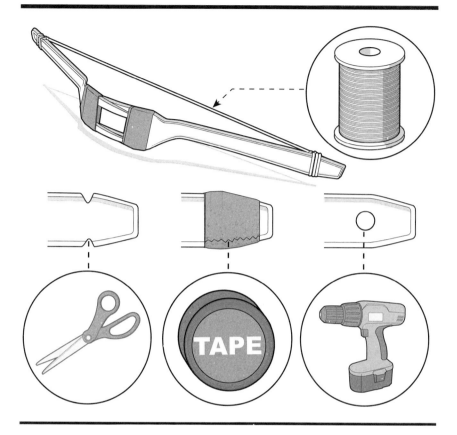

You have a few options to consider when mounting the bowstring to the fork assembly. The first option is to tie the string with a simple double knot around the end of each plastic handle, with the bowstring tight. Trim any excess string. If you feel this is sufficiently sturdy, then proceed to step 3.

Other bowstring mounting options include: carefully cutting small notches at both ends of the bow to hold the string in place, adding tape around the knotted string, or slowly drilling two small holes at the ends of the fork handles.

Step 3

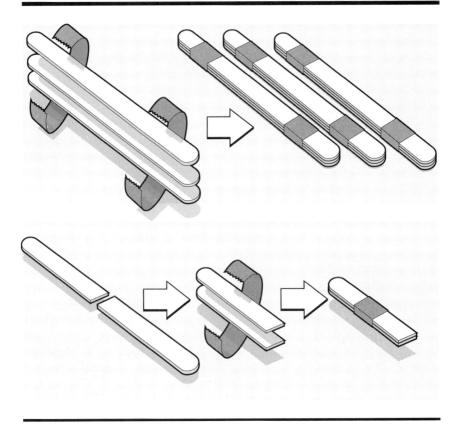

Take nine craft sticks and divide them into three piles, each three sticks high. Tape each stack together by placing the tape ¾ inch from each end (top image).

Next, with large scissors, carefully cut one craft stick in half. Then stack both halves on top of one another and secure the stack with tape (bottom image).

Step 4

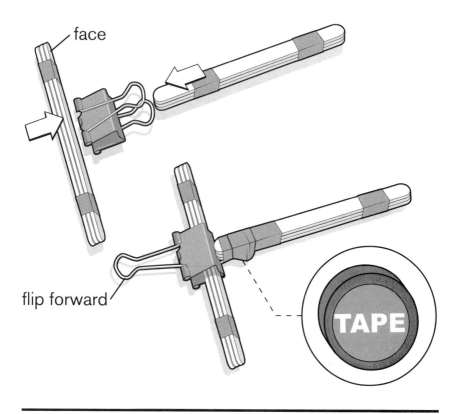

face

flip forward

TAPE

Clamp a medium binder clip onto the face of one of the craft stick bundles, positioning the binder clip in the center of the bundle. Either a 32 mm or 25 mm medium binder clip will work because its housing is the ideal size to fit around the face of the craft stick bundle.

Slide a second bundle of craft sticks between the attached binder clip metal handles. Flip the top handle forward toward the opening, and then tightly tape *only the bottom handle* onto the bundle of the craft sticks.

Step 5

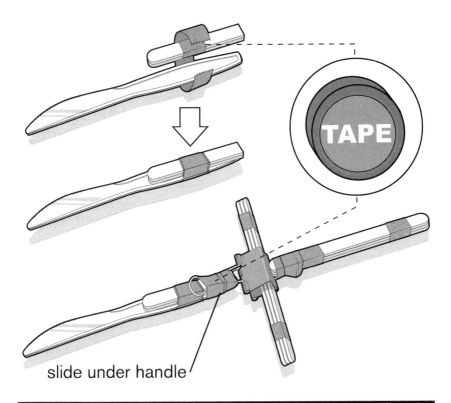

slide under handle

Continue constructing the frame by taping the half–craft stick bundle to the end of the plastic knife handle as shown.

With the binder clip handle still flipped forward from step 4, slide the knife assembly underneath the handle and tape it tightly in place.

Step 6

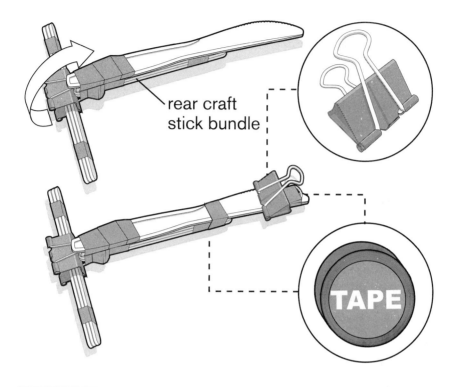

rear craft stick bundle

TAPE

Now flip the front binder clip assembly back to its original position. Once flipped, the rear craft stick bundle should be covered by the knife. Then, toward the cutting edge of the plastic knife, tape the knife and bundle together.

Secure a small binder clip to the back of the knife by taping around the bottom metal handle first and then adding additional tape inside the binder clip frame. The binder clip must still function properly after attached.

Step 7

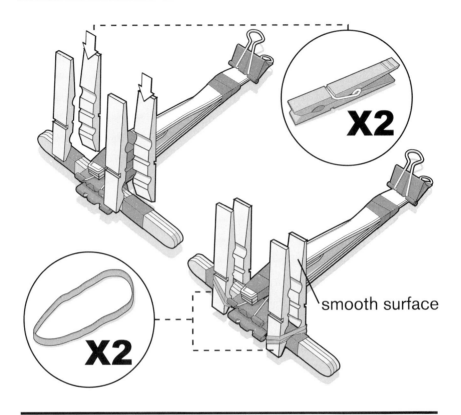

smooth surface

X2

X2

Disassemble two wooden clothespins by removing the metal springs between the prongs. Discard the metal springs.

Tight against the centered binder clip, secure the bottoms of both sets of clothespin prongs to the attached craft stick bundle with rubber bands, making sure the smooth surfaces face outward, as shown. Additional supports will be added to straighten these connections in the next steps.

Step 8

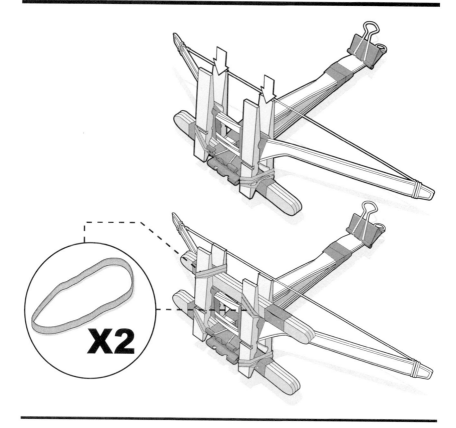

Slide the plastic fork assembly (bow) between the attached, upright clothespin prongs. The plastic fork window detail should be centered between the prongs, with the bowstring above the rear frame.

Once the bow is in place, slide the last bundle of craft sticks between the clothespin prongs, above the bow assembly. Use two rubber bands to hold the bundle in place by wrapping the bands around the prongs. Once attached, the prongs should be straightened to 90-degree angles and aligned with one another.

Step 9

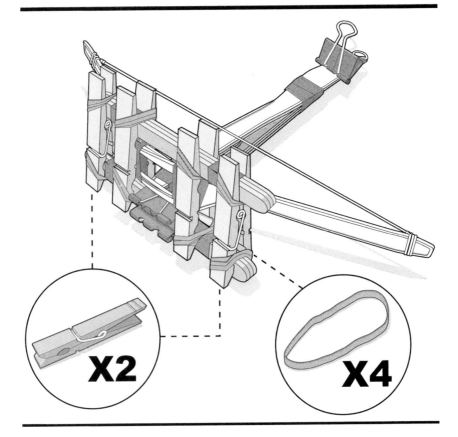

X2

X4

Affix the remaining two working clothespins to the ballista assembly by clamping them onto the lower bundle ends and then sliding the rear clothespin prongs around the upper bundle. Once aligned and at a 90-degree angle, use four rubber bands to fasten the clothespins to the frame.

Step 10

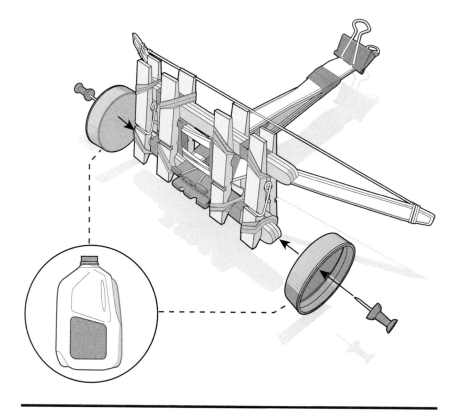

To add wheels to the ballista frame, center two pushpins through two plastic milk jug caps (or similar), then attach both wheels by pushing the pushpin point between the craft sticks in the lower bundle.

Step 11

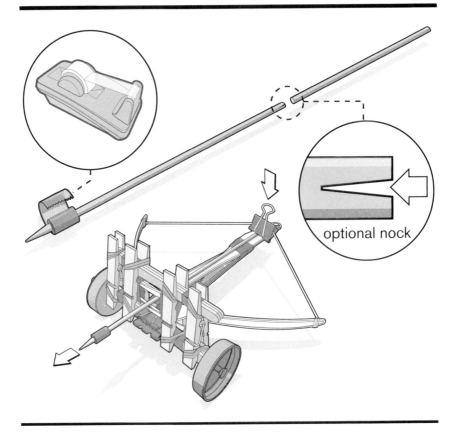

optional nock

You will arm this ballista with a pointed wooden skewer bolt (arrow). With the scissors, carefully shorten the skewer to 8 inches.

To increase the wooden bolt's accuracy, add weight to the point. Tightly wrap a 12-inch length of clear tape ½ inch from the pointed tip. You may need to decrease or increase the amount of tape until you arrive at the proper balance. If a wooden skewer is unavailable, try one of the other arrows used in this book.

To load, lock the bowstring into the attached binder clip, then load the wooden skewer through the ballista opening, with the blunt end resting in front of the loaded binder clip. To fire, carefully line up your target and release the clip. **Remember to use eye protection!** Bolts (arrows) can travel at a high velocity and have sharp points. **MiniWeapon projects are not meant for living targets**. Always stay clear of spectators and operate the bow in a controlled manner. Homemade weaponry can malfunction.

MOUNTED SIEGE BALLISTA

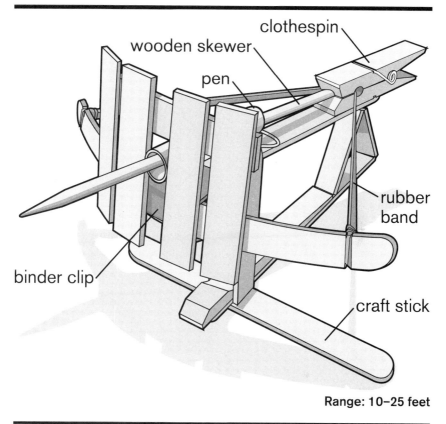

clothespin

wooden skewer

pen

rubber band

binder clip

craft stick

Range: 10–25 feet

The Mounted Siege Ballista is a scaled-down reconstruction of the Roman ballista that any medieval genius would appreciate. Loaded with a large wooden bolt, this wooden machine is designed to knock out targets at great distances.

Supplies

9 wooden craft sticks
3 rubber bands
2 wooden clothespins
1 medium binder clip (32 mm or 25 mm)
Duct tape
1 plastic ballpoint pen

Tools

Safety glasses
Hobby knife
2 2-peg-by-4-peg building blocks
Bowl of warm water
Hot glue gun
Large scissors (optional)

Ammo

1+ wooden skewers

Step 1

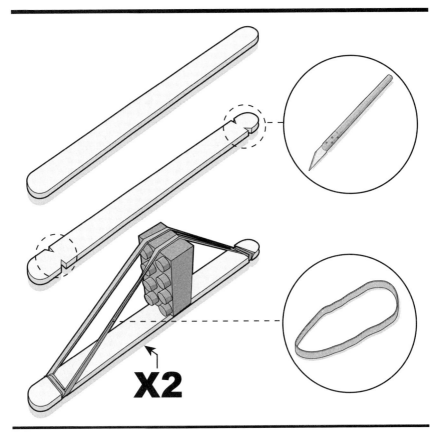

X2

With a hobby knife, carefully cut two small wedge details out of both ends of a craft stick, approximately ½ inch from the tips. Then loop one rubber band around both groups of notches. Once the band is in place, slide a 2-peg-by-4-peg building block (or something equivalent in size) upright, under the rubber band. Repeat this step with another craft stick for a total of two finished assemblies.

Step 2

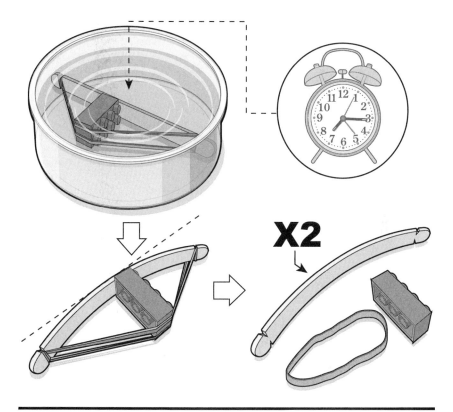

Wood bending with water is a fascinating process used by all types of woodworkers. You will be trying this craft by fully submerging both craft stick assemblies into a full bowl of warm water, saturating the wood to give them additional pliability. Soak the assemblies for 60 to 90 minutes before removing and bending them.

The rubber band should have assisted with the arcing process, but with the rubber band and block still attached, slowly increase the each arc with your fingers. Let the wet craft sticks dry for at least 30 minutes to allow the arc to set. Once dry, remove the rubber bands and toy blocks.

Step 3

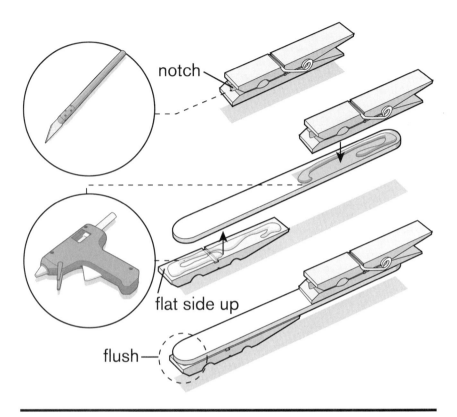

notch

flat side up

flush

With a hobby knife, carefully cut out a skewer-diameter-sized notch from the front prongs of a clothespin. The notch should be centered between the two prongs and only $1/16$ to $1/8$ inch deep. Once cut, test with a wooden skewer for a snug fit.

Next, hot glue the notched clothespin on to the end of a straight craft stick. The front of the clothespin housing should have approximately $1/2$ inch overlap when assembled.

Disassemble the second clothespin by removing the metal pin, and then hot glue one prong, flat side up, to the underside of the craft stick—opposite to the attached notched clothespin—with the prong and craft sick ends flush with one another.

Step 4

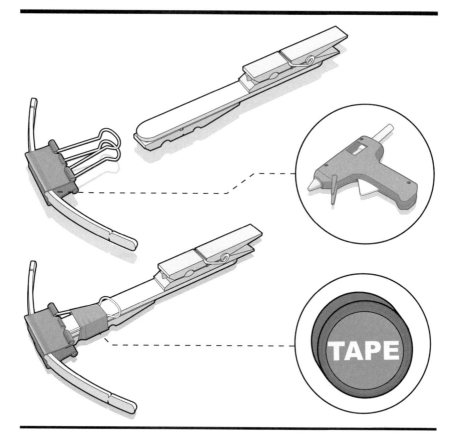

To construct the bow, hot glue and clamp one medium binder clip (32 mm or 25 mm) to the center of one of the bent craft sticks from step 2. When in place, the craft stick's arc should be bent toward the binder clip opening, as shown.

Slide the craft stick assembly between both metal handles of the binder clip—bottom prong end first—until the wooden end rests against the bent craft stick. Once aligned with the clip, join the assemblies by tightly wrapping tape around the metal handles.

Step 5

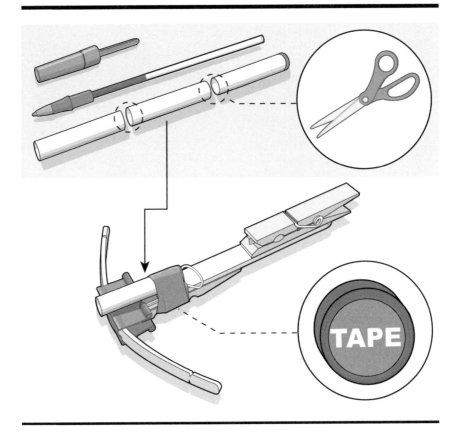

To create the ballista barrel, disassemble a plastic ballpoint pen into its various parts, removing the ink cartridge. Then cut the pen housing into three equal cylinders using the hobby knife or a pair of large scissors.

Tape one of the pen cylinders to the top of the frame assembly, with the pen housing protruding past the front binder clip slightly ($\frac{1}{16}$ to $\frac{1}{8}$ of an inch).

Step 6

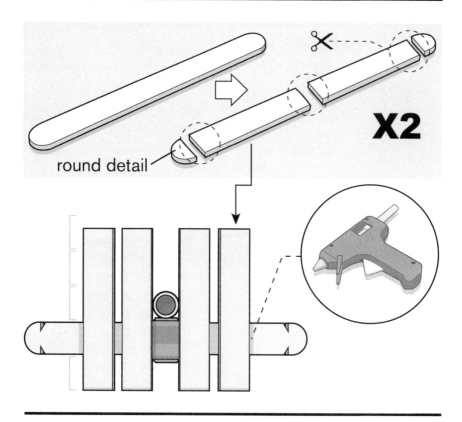

round detail

X2

To protect ballista operators during ancient battles, a wooden shield was fixed to various ballista designs, depending on the weapon's use or the culture. You will create an authentic shield by modifying two craft sticks and then attaching the various parts to the front of the frame.

To begin, cut the four round ends off two craft sticks. Then cut the rectanglular craft sticks in half to create four equal-length mini boards. If necessary, align and trim the boards to make the lengths equal.

Using the illustration as a guide, hot glue two craft stick halves to the attached binder clip bottom. With both halves spaced next to the pen housing barrel, position them asymmetrically, with ½ inch overhanging the bottom bent clip (bow). Then take the remaining two halves and hot glue them ⅛ inch from the first grouping.

Step 7

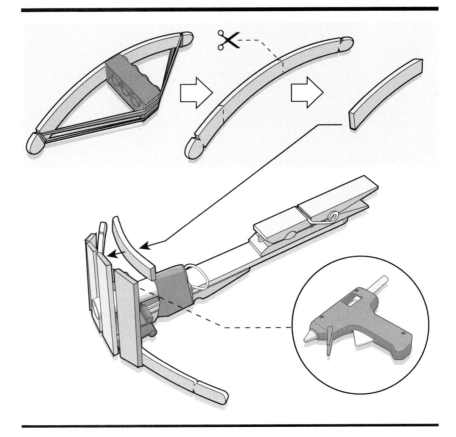

Add additional support behind the wooden shield by cutting the sec-ond bent craft stick to the same width as the front shield. Once it is cut to the correct length, carefully hot glue the curved section behind the four attached sticks, slightly above the pen barrel housing.

Step 8

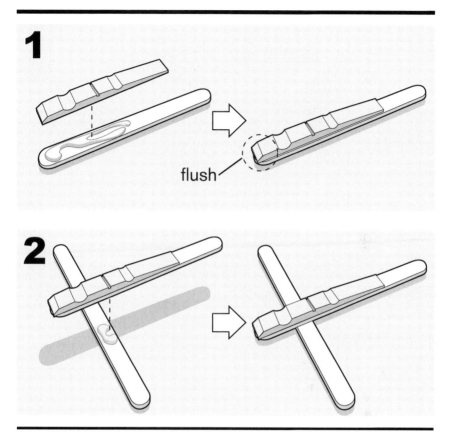

flush

The stand for this ballista is modeled after an authentic Roman ballista. To build it, glue the second clothespin prong from step 3, flat surface down, to the top of a craft stick. The prong and craft sticks ends should be flush (1).

Carefully hot glue this assembly to the top of a second craft stick, ¾ inch from the front of the assembly and at 90 degrees (2).

Step 9

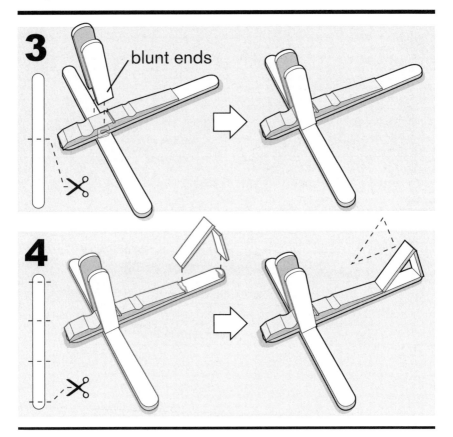

With the hobby knife or large scissors, carefully cut another craft stick in half. Then hot glue the blunt ends of both halves to the sides of the prong at a 90-degree angle, where the craft sticks intersect (3).

Next, cut both round ends off the last craft stick. Then cut the craft stick into three equal parts. Hot glue two of the three parts onto the back of the assembly, creating a triangle (4). This triangle will support the back of the ballista frame in the next step.

Step 10

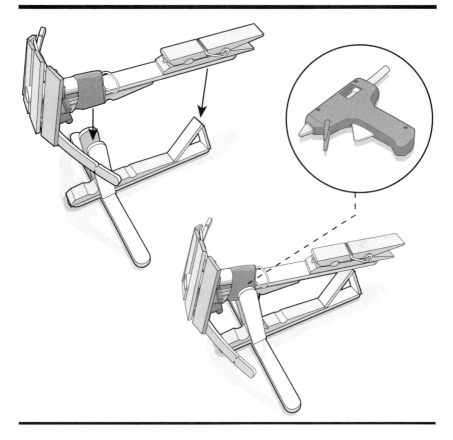

Combine both assemblies by placing the ballista frame between the two upright craft stick halves on the base. Slightly angle the ballista frame to rest on the rear triangle detail. Add hot glue to the two uprights and underside of the triangle detail to lock both assemblies together.

Step 11

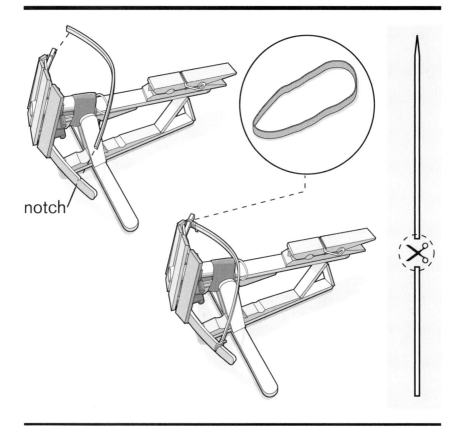

notch

For this ballista, you will be using a nontraditional rubber band bow-string. Cut open a rubber band and straighten it out. Then tie each end of the rubber band onto opposite ends of the bow limbs, using the notches to help hold the rubber band knot. The ballista is now complete.

A modified wooden skewer is suggested for the bolt (arrow). With the hobby knife or scissors, shorten the wooden skewer to a length of roughly 4½ inches. If a skewer is unavailable, see one of the other projects in this book for a substitute bolt.

To fire, lock the bowstring (rubber band) into the clothespin, then load the bolt through the pen barrel and into the clothespin notch. Release when ready. ***Remember to use eye protection!*** Bolts (arrows) can travel at a high velocity and have sharp points. ***Mini-Weapon projects are not meant for living targets***. Always stay clear of spectators and operate the bow in a controlled manner. Home-made weaponry can malfunction.

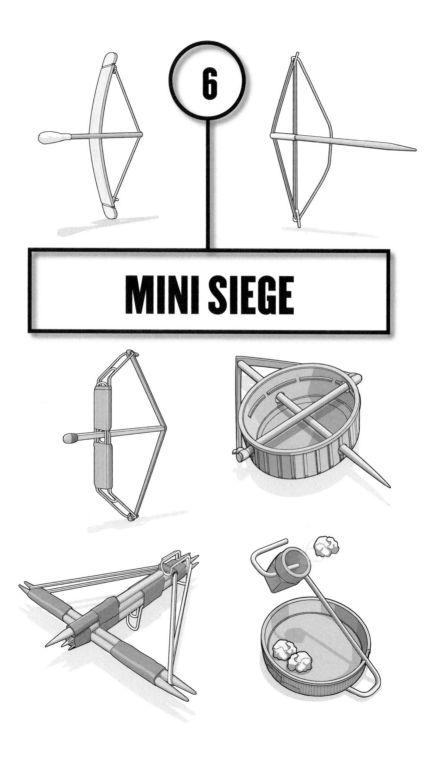

MINI SIEGE

CRAFT STICK MINI BOW

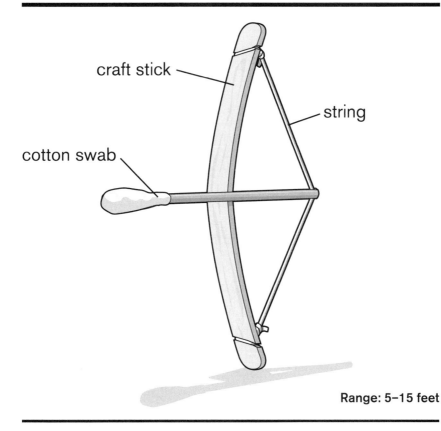

craft stick

string

cotton swab

Range: 5–15 feet

Master desktop archery with the Craft Stick Mini Bow, one of the most basic bow designs in the MiniWeapons arsenal. Constructed from a single craft stick, it's an excellent choice when outfitting the king's army or a bunch of merry men.

Supplies

1 craft stick
1 rubber band
String

2-peg-by-4-peg building block
Bowl of warm water
Scissors

Tools

Safety glasses
Hobby knife

Ammo

1+ plastic cotton swabs

Step 1

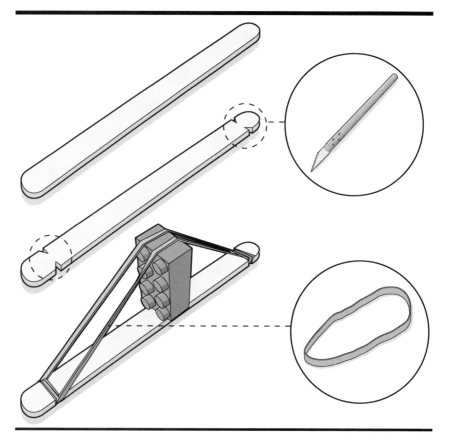

With a hobby knife, carefully cut two small wedge details out of both ends of a craft stick, approximately ½ inch from the tips. Then loop one rubber band around both groups of notches. Once the band is in place, slide a 2-peg-by-4-peg building block (or something equivalent in size) upright, under the rubber band.

Step 2

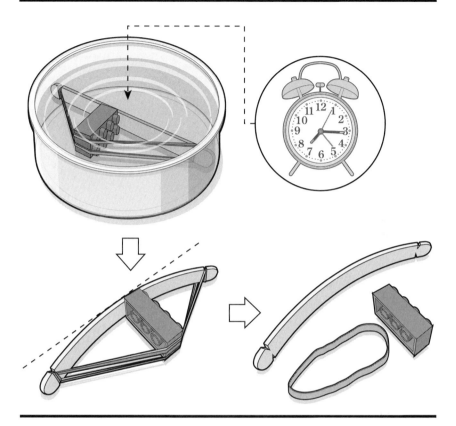

As in the Compound Crossbow (page 175) and Mounted Siege Ballista (page 213) projects, you will be sampling the art of wood bending. Fully submerge the craft stick assembly in a bowl of warm water, saturating the wood to give it additional pliability. Soak the assembly for 60 to 90 minutes before removing it.

The rubber band should have assisted with the arcing process, but with the rubber band and block still attached, slowly increase the arc with your fingers. Let the wet and bent craft stick dry for at least 30 minutes to allow the arc to set. Once the craft stick is dry, remove the rubber band and block.

Step 3

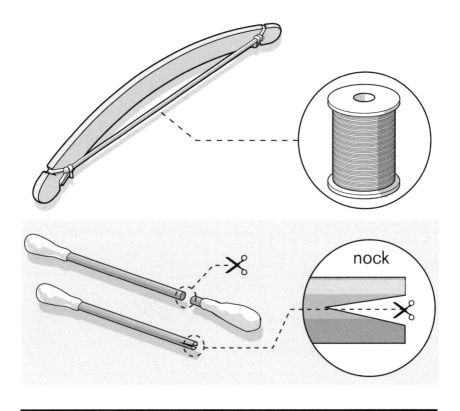

You will be using a traditional bowstring for this MiniWeapon; however, a rubber band could be substituted for similar results. Tie the string to each end of the bow, using the notch to secure the knots in place. The finished bowstring should be tight, with tension.

The soft-tipped arrow will be constructed out of one cotton swab. With scissors, remove one cotton end, then add a small nock to the cut end. Use this small notch to keep the cotton swab arrow in place on the bowstring as the craft stick bow is being drawn.

Remember to use eye protection! Arrows can travel at a high velocity and have sharp points. ***MiniWeapon projects are not meant for living targets***. Always stay clear of spectators and operate the bow in a controlled manner. Homemade weaponry can malfunction.

PAPER CLIP BOW

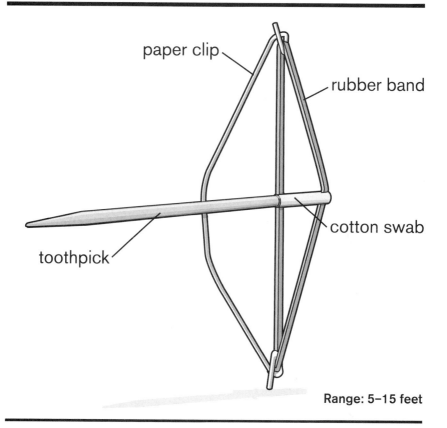

paper clip

rubber band

cotton swab

toothpick

Range: 5–15 feet

Transform a boring paper clip into a menacing projectile weapon with this Paper Clip Bow. Designed to be built quickly, this simple bow is perfect for a friendly game of target practice using the print-out targets in the back of the book (page 265).

Supplies

1 large paper clip
1 rubber band

Ammo

1+ round wooden toothpicks
1+ plastic cotton swabs

Tools

Safety glasses
Needle-nose pliers
Scissors

Step 1

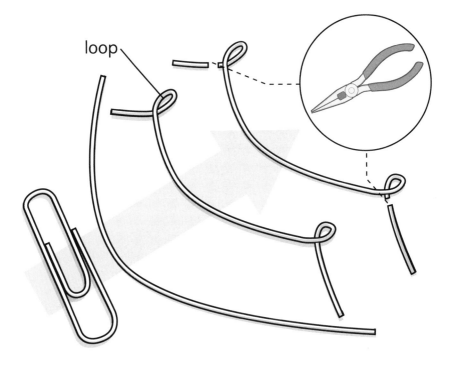

loop

Scrounge around your kingdom for one large paper clip. Then, using brute strength, bend and strengthen the paper clip housing to reasonable a slight "bow-shaped" arc as shown (second from left).

Once arced, use needle-nose pliers to loop both ends of the paper clip. With the pliers, trim off any excess paper clip that extends past the loop.

Step 2

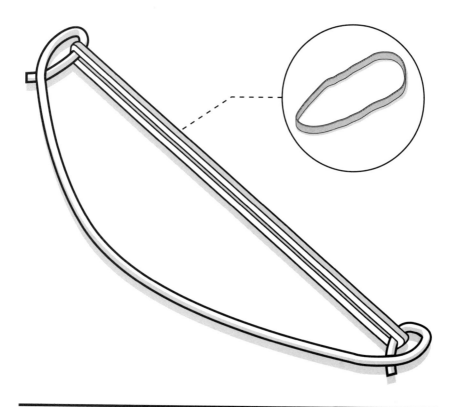

Hook one rubber band around both paper clip loops. (A string can be substituted.) The bowstring is now complete.

Step 3

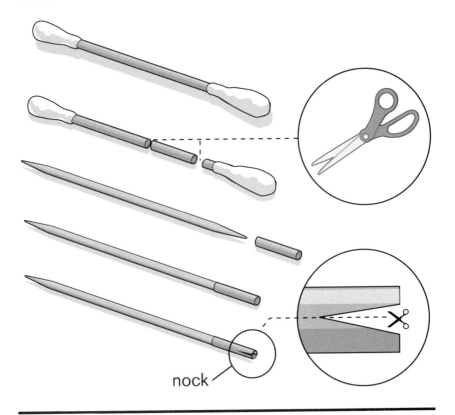

nock

Each arrow will be constructed from one plastic cotton swab and one round wooden toothpick. Use scissors to cut one cotton end off a plastic cotton swab. Then cut another ¼ inch section off the cotton swab handle. Slide the removed ¼-inch cylinder onto the round toothpick. Once in place, cut a small notch out of the back of the cylinder for the arrow nock.

Use this small notch to keep the toothpick arrow in place on the bowstring (rubber band) as the Paper Clip Bow is being drawn. **Remember to use eye protection!** Arrows can travel at a high velocity and have sharp points. **MiniWeapon projects are not meant for living targets**. Always stay clear of spectators and operate the bow in a controlled manner. Homemade weaponry can malfunction.

TRI-CLIP BOW

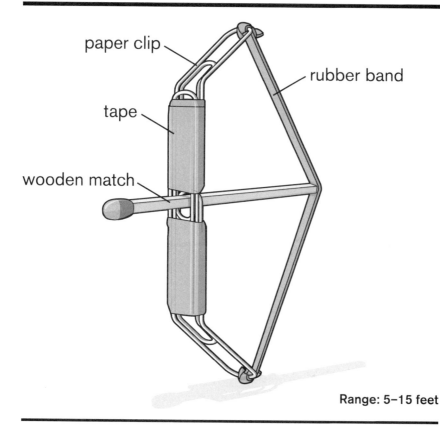

paper clip

rubber band

tape

wooden match

Range: 5–15 feet

The Tri-Clip Bow is cleverly constructed out of three large paper clips, fused together into a kick-butt design junior archers have been waiting for! Powered by a nontraditional elastic bowstring, with customizable matchstick darts, it's a great tool for fighting off bandits.

Supplies

3 large paper clips
Duct tape
1 rubber band

Ammo

1+ wooden matches or toothpicks
Piece of scrap paper (optional)

Tools

Safety glasses
Wire cutters or pliers
Scissors
Hobby knife (optional)

Tri-Clip Bow

Step 1

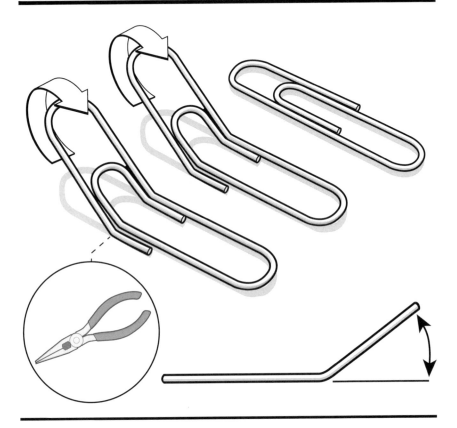

Using wire cutters or pliers, bend two large paper clips at the center by 45 degrees, as shown. Do not bend the third clip.

Step 2

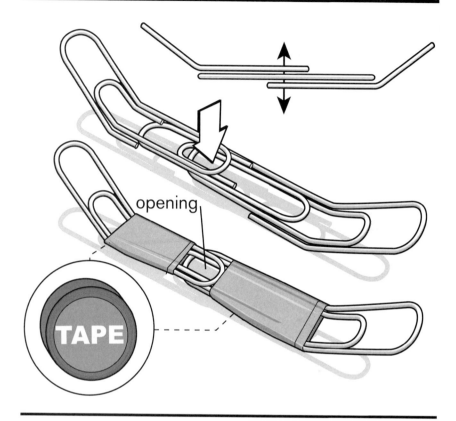

opening

TAPE

Stack all three paper clips, with the straight paper clip sandwiched in the middle and the two bent clips mirroring one another, as shown. Keep the center opening of the combined clips unobstructed. Tightly tape all three paper clips together, but *do not* cover the center opening of the assembly.

Step 3

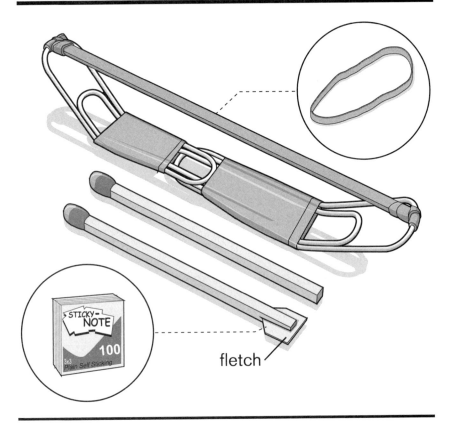

fletch

With scissors, cut open a rubber band and straighten it out. With tension, tie each end of the rubber band onto opposite ends of the bent clips. Complete the step by cutting off any excess rubber band. The bow is complete.

For arrows, wooden matches work great. They can be customized by adding a small fletch (fin) to the back of them. To do this, first use a hobby knife to cut a small notch into the back of the wooden housing. Then slide a small-sized piece of scrap paper into the notch with cut slants or fletch details. If a wooden match is unavailable, try toothpicks, cotton swabs, or any of the arrows found in this book's projects.

Remember to use eye protection! Arrows can travel at a high velocity and have sharp points. *MiniWeapon projects are not meant for living targets.* Always stay clear of spectators and operate the bow in a controlled manner. Homemade weaponry can malfunction.

BOTTLE CAP CROSSBOW

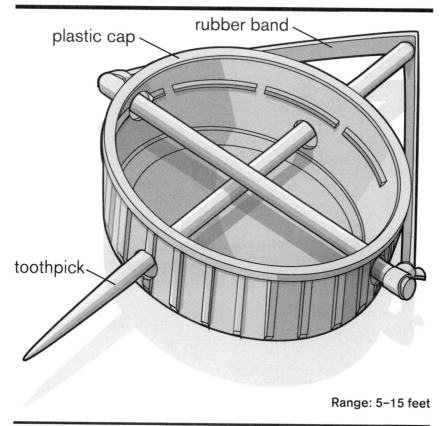

plastic cap

rubber band

toothpick

Range: 5–15 feet

Infiltrate the castle with the concealable Bottle Cap Crossbow! Constructed from a single plastic housing, its nontraditional design integrates both the frame and stock into one durable little devil. Close-range targets beware—this siege weapon is thirsty for a bull's-eye.

Supplies

1 plastic bottle cap
1 toothpick
1 rubber band

Tools

Safety glasses
Hobby knife
Scissors

Ammo

1+ toothpicks

Step 1

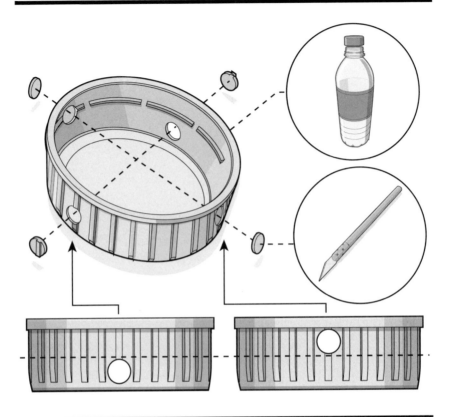

Carefully use a hobby knife to cut four toothpick-diameter-sized holes in a plastic bottle cap. Each pair of holes should be perpendicular to the other, but the pairs should be on separate planes. Use the side views in the bottom illustrations as a guide. If space is limited, find a larger plastic cap.

Step 2

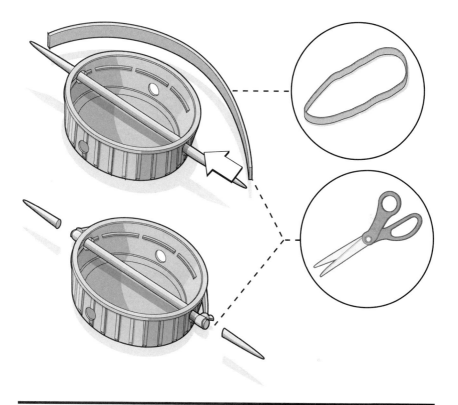

Slide one toothpick through the top two cut holes.

Next, attach the bowstring to the toothpick. With scissors, cut open the rubber band and straighten it out. With tension, tie each end of the rubber band onto opposite ends of the toothpick. Complete this step by cutting off any excess rubber band as well as the pointed tips of the toothpick.

Step 3

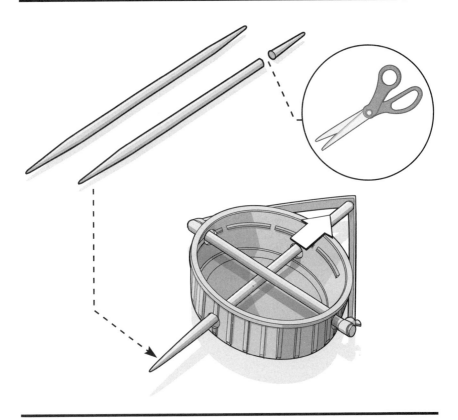

For the crossbow bolt, use scissors to remove one pointed end of the toothpick. Slide the blunt end of the toothpick into the cap's bottom set of holes. Grasp both the rubber band and the blunt end of the toothpick and slowly pull back. Find your target and release. If a toothpick is unavailable, a shortened cotton swab can be substituted.

Remember to use eye protection! Bolts (arrows) can travel at a high velocity and have sharp points. *MiniWeapon projects are not meant for living targets*. Always stay clear of spectators and operate the bow in a controlled manner. Homemade weaponry can malfunction.

TOOTHPICK CROSSBOW

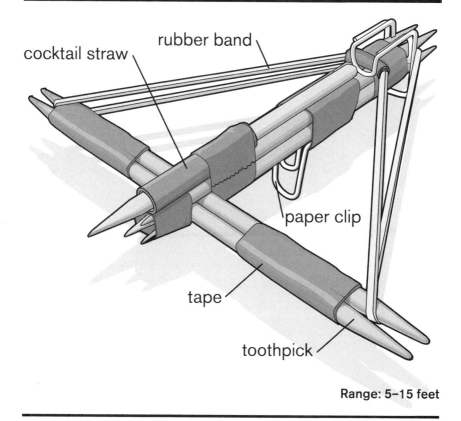

cocktail straw

rubber band

paper clip

tape

toothpick

Range: 5–15 feet

This pocket-sized crossbow is perfect for a miniature medieval battle. It features a working trigger mechanism mounted on the back, is easy to assemble, and is inconspicuous. Just don't let your guard down by revealing its impressive engineering and functionality—the enemy is always watching.

Supplies

6 round wooden toothpicks
Duct tape
1 cocktail straw
2 small paper clips
1 rubber band

Tools

Safety glasses
Scissors
Needle-nose pliers

Ammo

1+ toothpicks

Step 1

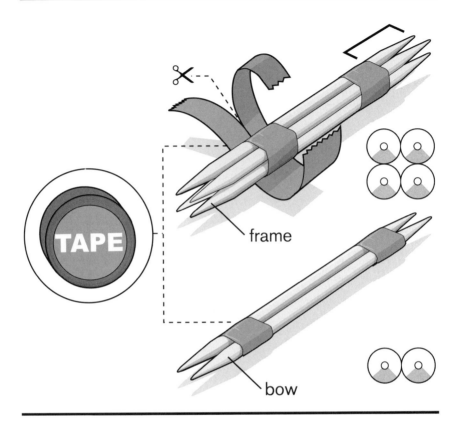

frame

bow

In this step, you will be constructing the frame and bow out of tooth-picks and tape. Prior to taping, cut the tape into ¼-inch-wide strips, because this mini frame is small.

Bundle four round wooden toothpicks to form a square and tightly tape the bundle ½ inch from both ends, as shown (top). This bundle of four is the frame.

For the second bundle, tape two toothpicks together, with tape ¼ inch from both ends (bottom). This bundle is the bow.

Step 2

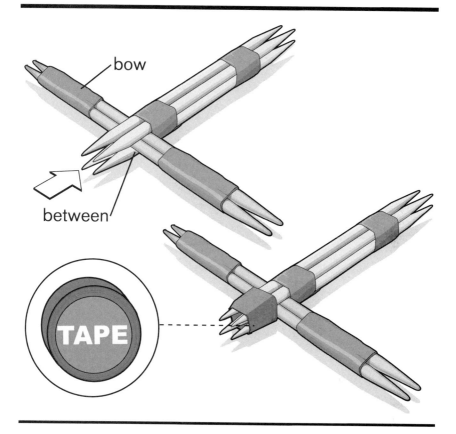

bow

between

TAPE

With the bow rotated 90 degrees, slide it between the frame, with two toothpicks on top and two beneath, until the bow rests against the tape of the frame. Center the bow and hold it in place by carefully cutting a piece of tape to the width of the front of the frame.

Step 3

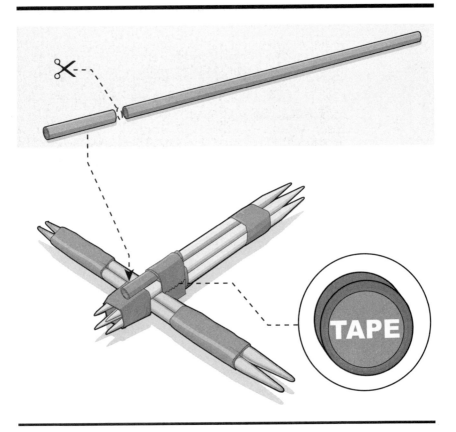

With scissors, cut ¾ inch off a cocktail straw (small drinking straw). Tape the smaller section to the top of the toothpick frame, above the bow intersection. The straw should be almost flush with the front of the toothpick assembly.

Step 4

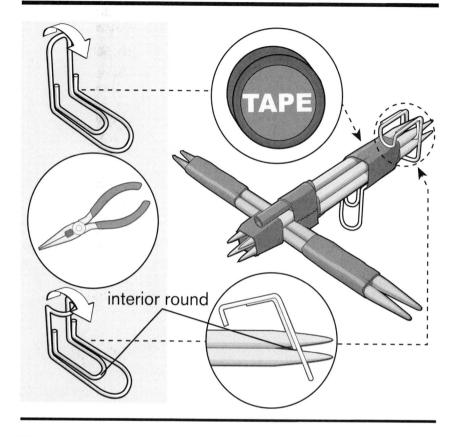

interior round

TAPE

To make the trigger mechanism and crossbow handle, use needle-nose pliers to bend two small paper clips. For the first paper clip (top left), bend one 90-degree angle at the center. Then tape this paper clip to the underside of the frame, 1 inch off the back of the assembly (opposite end from the bow).

For the second paper clip (bottom left), bend a 90-degree angle $^3/_8$ inch from one end. Place a second 90-degree angle at the center of the housing. Then slide this paper clip onto the rear of the frame, with the interior round detail between the two pairs of toothpicks. The outer round detail will rest below the frame. This paper clip will rest in place until a rubber band is added in step 5.

Step 5

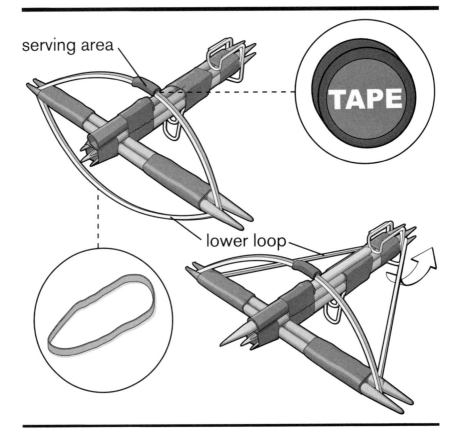

serving area

TAPE

lower loop

Loop one rubber band between the two tips of the bow. In the middle section of the upper side of the rubber band, wrap a small piece of tape, 1 inch long by ¼ inch wide. This will be the serving area, the part of the string where the bolt will rest when the crossbow is cocked.

Pull the lower loop of the rubber band toward the back and place it between the two pairs of toothpicks in the frame to hold in the second paper clip (trigger mechanism).

Step 6

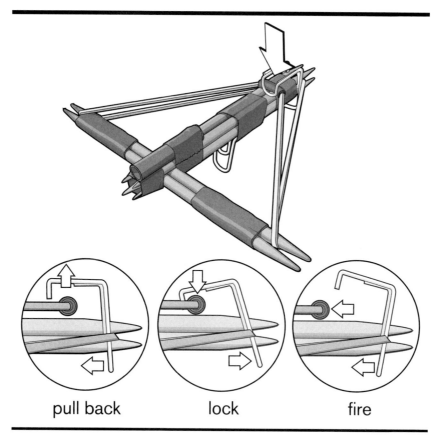

pull back   lock   fire

To test fire the crossbow, pull back the rubber band bowstring and lock the serving area (added tape) behind the 90-degree-angled paper clip. Then, from the underside, pull the bottom of the paper clip backward to keep the top of the paper clip locked onto the bowstring. When you are ready to fire, push the bottom of the paper clip forward to release the bowstring.

Step 7

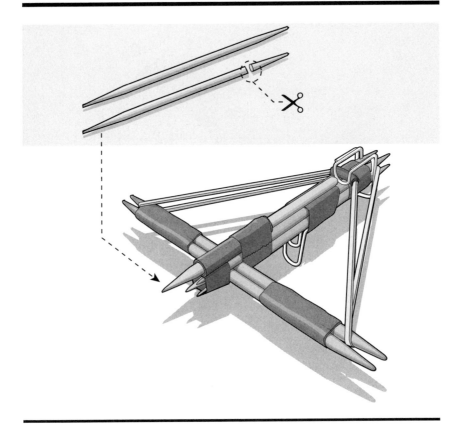

This crossbow will be armed with a toothpick bolt. With scissors, remove one pointed end of the toothpick. To fire, lock the bowstring into the paper clip, then load the blunt end of the wooden bolt into the straw barrel, with the blunt end just in front of the locked trigger. Once locked and loaded, fire at will. If a toothpick is unavailable, a cut cotton swab can be substituted, but only if a larger tube is used for the barrel.

Remember to use eye protection! Bolts (arrows) can travel at a high velocity and have sharp points. **MiniWeapon projects are not meant for living targets**. Always stay clear of spectators and operate the bow in a controlled manner. Homemade weaponry can malfunction.

CLIP AND CAP CATAPULT

spitball

paper clip

pencil top eraser

milk jug cap

tape

Range: 5–15 feet

Hurl candies and spitballs with the Clip and Cap Catapult, a micro-launcher designed to stop invading armies or peg unsuspecting targets from a safe distance. It can be assembled in minutes, and its base is integrated with a sweet ammo holder to support any relentless onslaught!

Supplies

1 pencil top eraser
1 large paper clip
1 plastic milk jug cap (or similar)
Duct tape

Tools

Safety glasses
Large scissors or hobby knife

Ammo

1+ spitballs, soft candies, or
   pencil erasers

Step 1

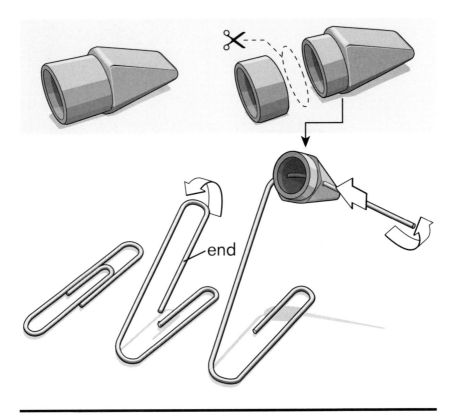

end

With large scissors or a hobby knife, carefully cut ³/₈ inch off the connection end of a pencil top eraser. The tip will be used as the catapult
basket. Save the removed eraser for ammo.

Bend the outer frame of one large paper clip at a 45-degree angle.
Then, on the 45-degree angle, bend the end of the clip 90 degrees
upward.

Slide the cut pencil top eraser through the 90-degree paper clip
rod as shown.

Step 2

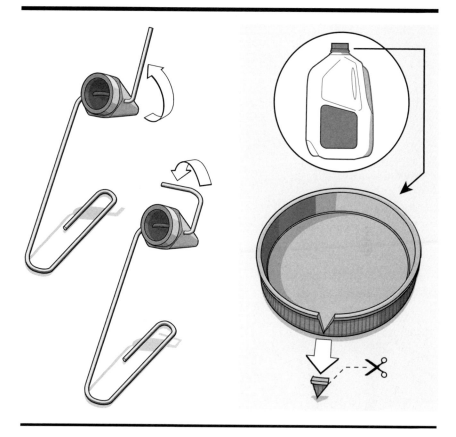

The paper clip will need to be bent two more times to create the pull-down release. Start by bending the rod protruding out of the eraser 90 degrees upward, then bend the same end another 90 degrees, over the eraser.

The base will be built out of a plastic milk jug cap (or similar). With the large scissors or hobby knife, carefully cut a small ⅛-inch wedge out of the outer ring of the milk jug cap.

Step 3

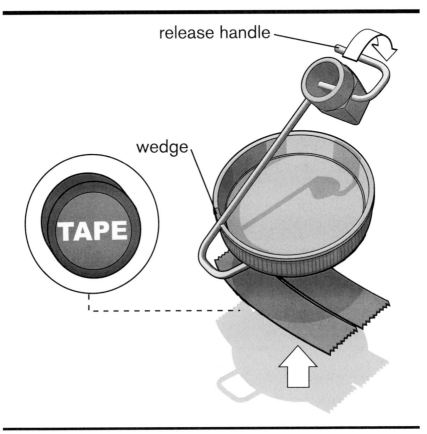

release handle

wedge

TAPE

Slide the lower brace of the swing arm assembly under the upside-down milk jug cap, with the 45-degree swing arm resting in the plastic cap's wedge. Join the parts by placing tape on the underside of the plastic cap, as shown. The Clip and Cap Catapult is now complete.

To fire, load the catapult basket (pencil top eraser), hold the lower frame for support, then pull back the release handle on the swing arm. Bombs away!

Remember to use eye protection! Never aim this catapult at a living target and use only safe ammunition. Spitballs, soft mints, and pencil erasers work nicely.

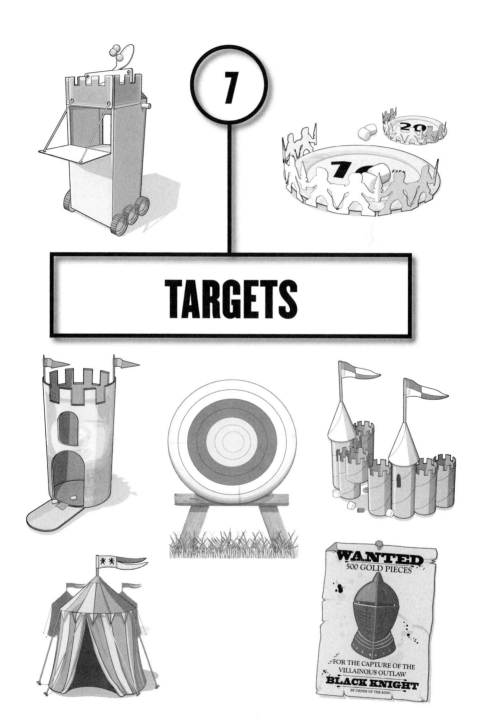

7

TARGETS

WANTED
500 GOLD PIECES

FOR THE CAPTURE OF THE
VILLAINOUS OUTLAW
BLACK KNIGHT
BY ORDER OF THE KING

CARTON SIEGE TOWER

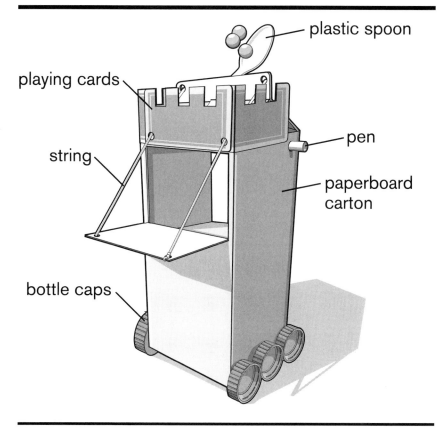

plastic spoon

playing cards

pen

string

paperboard carton

bottle caps

Stand your ground! With the catapults pounding the castle walls, it's only a matter of time until the siege tower rolls in for the invasion. Hone your defense by targeting the Carton Siege Tower. Built with common household items, it's a great addition to the battlefield. But beware—this siege tower is capable of firing back!

Supplies

1 paperboard carton (59 oz.)
3 playing cards
String
1 plastic ballpoint pen housing
1 plastic spoon
4–6 plastic bottle caps

Tools

Scissors or hobby knife
Single-hole punch
Hot glue gun

Step 1

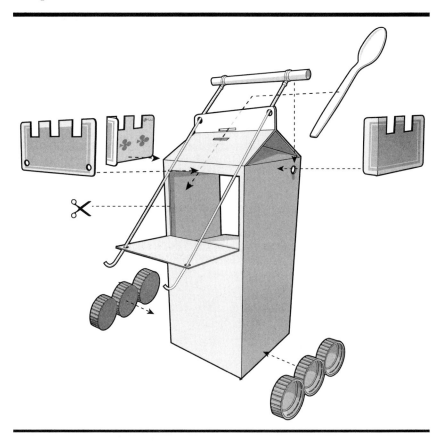

Start with a rinsed-out and dried paperboard (cardboard) carton (59 oz.). Cut out three sides of the front and rear doors as shown. Fold the front flap 90 degrees out; fold the rear flap 90 degrees into the carton.

For the top castle detail, cut three playing cards to form the iconic castle wall tops. Use a single-hole punch to cut two holes on the corners of the front door, bottom ends of one playing card, top of carton, and side rear of carton as illustrated. Then glue the cards to the carton top, bending the two unpunched cards around the sides.

Add string to raise and lower the front door. Tie the ends of the string to the front door, then run the string through the fixed playing card, through the top of the carton, and around a pen housing. Insert the pen housing into the holes in the sides of the box.

Use scissors or a hobby knife to cut two slits on the top of the carton to place one plastic spoon for firing. Then glue two or four plastic bottle caps to the bottom of the carton to simulate wheels.

ATTACKING ARMY

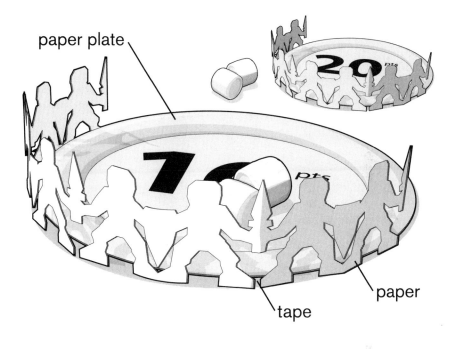

paper plate

paper

tape

Some competitive catapulting could do this kingdom some good! Master your medieval arsenal by creating several Attacking Army targets. With just some copy paper and a few paper plates, you can quickly fill the floor with a barbaric horde. To keep score, add point values to each paper plate; if the ammo hits or lands on that target, you receive that many points!

Supplies

1+ paper plates
1+ sheets of copy paper (8½ inches by 11 inches)
Tape (any kind)

Tools

Scissors
Art marker

Step 1

From a standard-sized sheet of copy paper, cut one long strip roughly 3 inches in width. Draw an outline of an armed soldier at one end of the strip, using the illustration for reference, where the paper's width is the height of the soldier. Add a sword or shield for a fun medieval look.

After the outline has been drawn, fold the paper accordion style, making sure the width of the folds is the same as the width of the drawn knight. The more paper folds, the more knights.

While the paper is folded, carefully use scissors to cut out the knights, leaving the folds intact. Unfold the paper to reveal a chain of knights.

Tape the chain of knights to the front of a paper plate. Use an art marker to add the point value of the target. Repeat these steps to create more targets.

TOWER OF OATMEAL

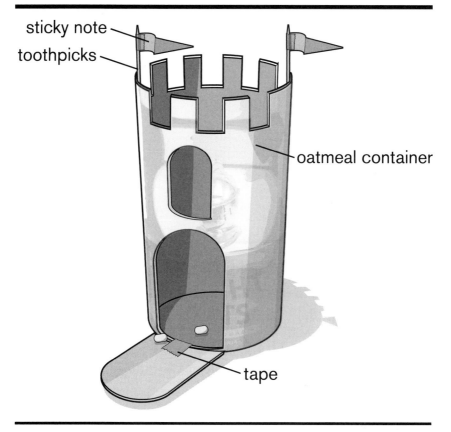

sticky note

toothpicks

oatmeal container

tape

During a battle, assaulting the high ground definitely provides a tactical challenge, since height increases the range of the enemy archers. You can practice siege techniques by building a Tower of Oatmeal. With a few critically placed hits, the mighty tower will fall and victory will be within reach.

Supplies

1 oatmeal container
1 square sticky note (3 inches by 3 inches) or sheet of paper
Tape (any kind)
2 toothpicks

Tools

Scissors or hobby knife

Step 1

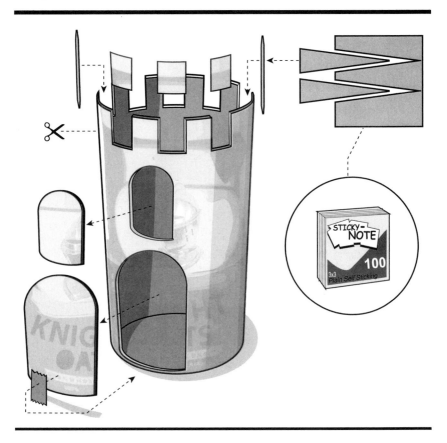

Remove the lid from a large, cylindrical oatmeal container. Use scissors or a hobby knife to create the iconic castle wall detail on the top of the container. Use the same tool to remove the window and drawbridge, as shown. Tape the drawbridge back onto the base of the tower.

   Then cut out two small, triangular castle flags from a sticky note or sheet of paper. Tape each flag onto a toothpick and then tape the toothpicks onto the top of the tower. Customize your target for added fun—for instance, by adding figures inside the tower.

STORM THE CASTLE

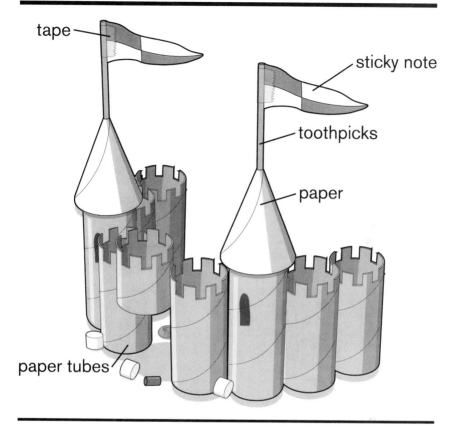

tape

sticky note

toothpicks

paper

paper tubes

What is a castle siege without a castle? With cardboard tubes, create a custom fortress fit for a king that is limited only by your imagination. Test this stronghold against any number of MiniWeapon projects to see if it can withstand the mighty fist of ingenuity and creativity.

Supplies

9+ toilet paper tubes (or similar)
Masking tape or clear tape
1 sheet of copy paper (8½ inches by 11 inches)
1 square sticky note (3 inches by 3 inches)
2 toothpicks

Tools

Scissors or hobby knife
Art markers

Step 1

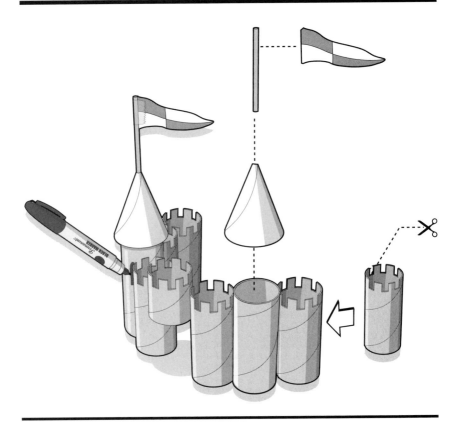

Round up nine toilet paper tubes or similar for the construction of the castle walls. Cut one end of seven tubes with the castle wall detail (embrasure) using scissors or a hobby knife. Then cut one of these tubes in half to create the castle entryway.

Following the illustration, tape the tubes together to resemble a castle wall. The half tube should be mounted between the two center tubes to create a door frame, as shown.

To create two pointed towers, roll two paper cones and then cut the cones to length. The diameter of the cones should be slightly larger than the cardboard tubes' diameter. Tape the cones onto the two uncut tubes.

Use markers to add windows or other details to the towers. Miniature flags can be cut out of square sticky notes; tape the banners to toothpicks and the finished flags to the paper cone towers. Have fun with it.

ARCHERY TARGET

Competitor_____ Date_____

Competitor Signature_____

Use a copy machine to make multiples and enlarge.

ENEMY CAMP

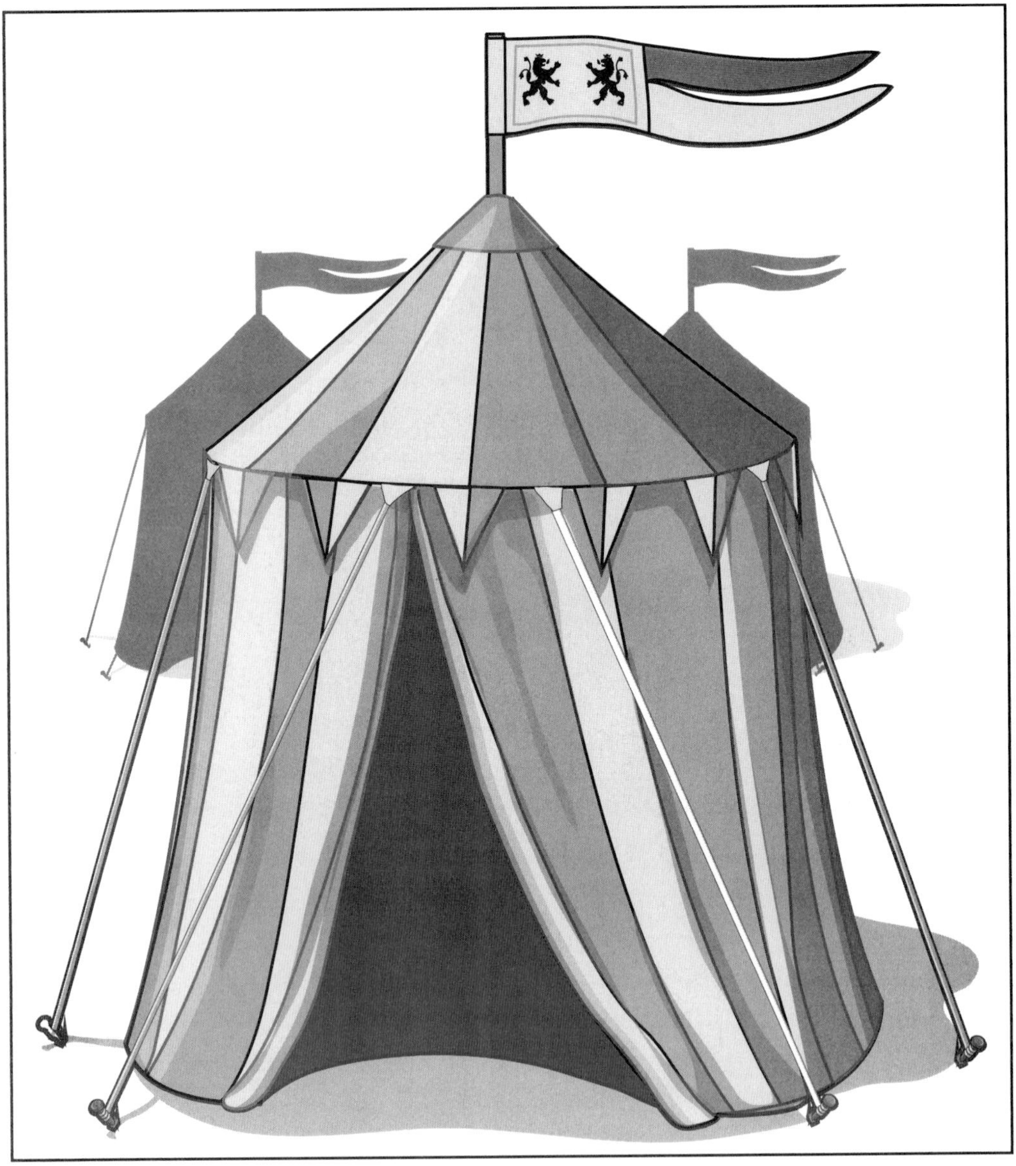

Competitor_____ Date_____

Competitor Signature_____

Use a copy machine to make multiples and enlarge.

WANTED POSTER

Competitor_____ Date_____

Competitor Signature_____

Use a copy machine to make multiples and enlarge.

For more information and free
downloadable targets, please visit:

MINIWEAPONSBOOK.COM

ALSO FROM CHICAGO REVIEW PRESS

MiniWeapons of Mass Destruction

Build Implements of Spitball Warfare

John Austin

978-1-55652-953-5
$16.95 (CAN $18.95)

We've come a long way from the Peashooter Era! Using items that can be found in the modern junk drawer, troublemakers of all stripes have the components they need to assemble an impressive arsenal of miniaturized weaponry.

MiniWeapons of Mass Destruction provides fully illustrated step-by-step instructions for building 35 projects, including:

- ➲ Clothespin Catapult
- ➲ Matchbox Bomb
- ➲ Shoelace Darts
- ➲ Paper-Clip Trebuchet
- ➲ Tube Launcher
- ➲ Clip Crossbow
- ➲ Coin Shooter
- ➲ Hanger Slingshot
- ➲ Ping-Pong Zooka
- ➲ And more!

And for those who are more MacGyver than marksman, *MiniWeapons* also includes target designs, from aliens to zombies, for practice in defending their personal space.

MiniWeapons of Mass Destruction 2

Build a Secret Agent Arsenal

John Austin

978-1-56976-716-0
$16.95 (CAN $18.95)

If you're a budding spy, what better way to conceal your clandestine activities than to miniaturize your secret agent arsenal? *MiniWeapons of Mass Destruction 2* provides fully illustrated step-by-step instructions for building 30 different spy weapons and surveillance tools, including:

- Paper Dart Watch
- Rubber Band Derringer
- Pushpin Dart
- Toothpaste Periscope
- Bionic Ear

- Pen Blowgun
- Mint Tin Catapult
- Cotton Swab .38 Special
- Paper Throwing Star
- And more!

Once you've assembled your weaponry, the author provides a number of ideas on how to hide your stash—inside a deck of cards, a false-bottom soda bottle, or a cereal box briefcase—and targets for practicing your spycraft, including a flip-down firing range, a fake security camera, and sharks with laser beams.

MiniWeapons of Mass Destruction Targets

100+ Tear-Out Targets, Plus 5 New MiniWeapons

John Austin

978-1-61374-013-2
$9.95 (CAN $10.95)

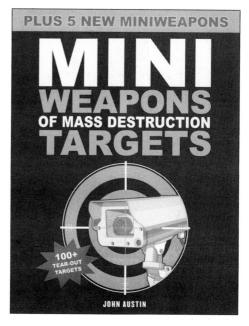

The key to becoming an accomplished marksman is to practice, practice, practice. *MiniWeapons of Mass Destruction Targets* contains more than 100 tear-out targets to develop your skills. The targets are divided into three themes—Basic, Secret Agent, and Dark Ages—with a variety of gameplay scenarios. Blast the lock off a chained door, knock down a castle gate, compete in a game of Around the World, or shoot several miniature targets at various locations. Rules on the back of each target describe basic and advanced play.

In addition to the 100+ targets, MiniWeapons master John Austin provides instructions for building five new MiniWeapons perfect for target shooting:

- ➲ Paper Pick Blow Gun
- ➲ Spitball Shooter with Clip
- ➲ Toothpick Tape Dart
- ➲ Semiautomatic Pen Pistol
- ➲ Pen Cap Dart

Safety instructions are also included, as well as a guide to setting up an in-house firing range that will protect walls and furniture.

So Now You're a Zombie

A Handbook for the Newly Undead

John Austin

978-1-56976-342-1
$14.95 (CAN $16.95)

Zombies know that being undead can be disorienting. Your arms and other appendages tend to rot and fall off. It's difficult to communicate with a vocabulary limited to moans and gurgles. And that smell! (Yes, it's *you*.) But most of all, you must constantly find and ingest human brains. *Braaaains!!!*

What's a reanimated corpse to do?

As the first handbook written specifically for the undead, *So Now You're a Zombie* explains how your new, putrid body works and what you need to survive in this zombiphobic world. Dozens of helpful diagrams outline attack strategies to secure your human prey, such as the Ghoul Reach, the Flanking Zeds, the Bite Hold, and the Aerial Fall. You'll learn how to successfully extract the living from boarded-up farmhouses and broken-down vehicles. Zombiologist John Austin even explores the upside of being a zombie. Gone are the burdens of employment, taxes, social networks, and basic hygiene, allowing you to focus on the simple necessities: the juicy gray matter found in the skulls of the living.

Practical Pyromaniac, The

Build Fire Tornadoes, One-Candlepower
Engines, Great Balls of Fire, and More
Incendiary Devices

William Gurstelle

978-1-56976-710-8
$16.95 (CAN $18.95)

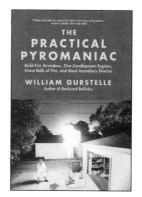

"What a fun, totally engrossing book! Gurstelle's projects—everything from a tiny single-candle engine to a flamethrower—are both easy to build and hard to resist. . . . Think of *The Practical Pyromaniac* as a cookbook for the budding scientist in each of us." —James Meigs, editor in chief of *Popular Mechanics*

The Practical Pyromaniac combines science, history, and do-it-yourself pyrotechnics to explain humankind's most useful and paradoxical tool: fire. William Gurstelle, frequent contributor to *Popular Mechanics* and *Make* magazine, presents dozens of projects with instructions, diagrams, photos, and links to video demonstrations that enable people of all ages (including young enthusiasts with proper supervision) to explore and safely play with fire.

Backyard Ballistics, 2nd edition

Build Potato Cannons, Paper Match Rockets,
Cincinnati Fire Kites, Tennis Ball Mortars, and
More Dynamite Devices

William Gurstelle

978-1-61374-064-4
$16.95 (CAN $18.95)

"How is it possible not to love a book with chapter titles like 'Back Porch Rocketry' and 'Greek Fire and the Catapult'? I devoured this prodigious account of all things explosive." —Homer Hickam, author of *Rocket Boys*

This bestselling guide has been expanded and updated, enabling ordinary folks to construct even more exciting ballistic devices in their garage or basement workshops than ever before. Clear instructions, diagrams, and photographs show how to build projects ranging from the simple—a match-powered rocket—to the more complex—a tabletop catapult—to the classic—the infamous potato cannon—to the offbeat—the Cincinnati fire kite. Four spectacular projects have been added to the fun arsenal: the spud-zooka, the powder keg, the electromagnetic pipe gun, and the sublimator.

Build Greek ballistae, Roman onagers, English Trebuchets, and more ancient artillery

The Art of the Catapult

Build Greek Ballistae, Roman Onagers, English Trebuchets, and More Ancient Artillery

William Gurstelle

978-1-55652-526-1
$16.95 (CAN $18.95)

"This book is a hoot . . . the modern version of *Fun for Boys* and *Harper's Electricity for Boys*."
—*Natural History*

Whether playing at defending their own castle or simply chucking pumpkins over a fence, wannabe marauders and tinkerers will become fast acquainted with Ludgar, the War Wolf, Ill Neighbor, Cabulus, and the Wild Donkey— ancient artillery devices known commonly as catapults. Instructions and diagrams illustrate how to build seven authentic, working model catapults, including an early Greek ballista, a Roman onager, and the apex of catapult technology, the English trebuchet.

Soda-Pop Rockets

20 Sensational Rockets to Make from Plastic Bottles

Paul Jarvis

978-1-55652-960-3
$16.95 (CAN $18.95)

Anyone can recycle a plastic bottle by tossing it into a bin, but it takes a bit of skill to propel it into a bin from 500 feet away. This fun guide features 20 different easy-to-launch rockets that can be built from discarded plastic drink bottles. After learning how to construct and launch a basic model, you'll find new ways to modify and improve your designs. Clear, step-by-step instructions with full-color illustrations accompany each project, along with photographs of the author firing his creations into the sky.

Gonzo Gizmos

Projects & Devices to Channel Your Inner Geek

Simon Field

978-1-55652-520-9

$16.95 (CAN $18.95)

This book for workbench warriors and grown-up geeks features step-by-step instructions for building more than 30 fascinating devices. Detailed illustrations and diagrams explain how to construct a simple radio with a soldering iron, a few basic circuits, and three shiny pennies; how to create a rotary steam engine in just 15 minutes with a candle, a soda can, and a length of copper tubing; and how to use optics to roast a hot dog, using just a flexible plastic mirror, a wooden box, a little algebra, and a sunny day. Also included are experiments most science teachers probably never demonstrated, such as magnets that levitate in midair, metals that melt in hot water, a Van de Graaff generator made from a pair of empty soda cans, and lasers that transmit radio signals.

Return of Gonzo Gizmos

More Projects & Devices to Channel Your Inner Geek

Simon Field

978-1-55652-610-7

$16.95 (CAN $22.95)

This fresh collection of more than 20 science projects—from hydrogen fuel cells to computer-controlled radio transmitters—is perfect for the tireless tinkerer. Its innovative activities include taking detailed plant cell photographs through a microscope using a disposable camera; building a rocket engine out of aluminum foil, paper clips, and kitchen matches; and constructing a geodesic dome out of gumdrops and barbecue skewers. Most of the devices can be built using common household products or components available at hardware or electronic stores, and each experiment contains illustrated step-by-step instructions with photographs and diagrams that make construction easy.

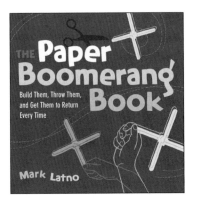

The Paper Boomerang Book

Build Them, Throw Them, and Get Them to Return Every Time

Mark Latno

978-1-56976-282-0
$12.95 (CAN $13.95)

The Paper Boomerang Book is the first-of-its-kind guide to this fascinating toy. Boomerang expert Mark Latno will tell you how to build, perfect, and troubleshoot your own model. Once you've mastered the basic throw, return, and catch, it's on to more impressive tricks—the Over-the-Shoulder Throw, the Boomerang Juggle, the Under-the-Leg Catch, and the dreaded Double-Handed, Backward, Double-Boomerang Throw. And best of all, you don't have to wait for a clear, sunny day to test your flyers—they can be flown indoors in almost any sized room, rain or shine.

The Flying Machine Book

Build and Launch 35 Rockets, Gliders, Helicopters, Boomerangs, and More

Bobby Mercer

978-1-61374-086-6
$14.95 (CAN $16.95)

Calling all future Amelia Earharts and Chuck Yeagers—there's more than one way to get off the ground! *The Flying Machine Book* will show you how to construct 35 easy-to-build and fun-to-fly contraptions that can be used indoors or out. Better still, each of these rockets, gliders, boomerangs, launchers, and helicopters can be made for little or no cost using recycled materials. Rubber bands, paper clips, straws, plastic bottles, and index cards can all be transformed into amazing, gravity-defying flyers, from Bottle Rockets to Grape Bazookas, Plastic Zippers to Maple Key Helicopters. Each project contains a materials list and detailed step-by-step instructions with photos, as well as an explanation of the science behind the flyer. Use this information to modify and improve your designs, or explain to your teacher why throwing a paper airplane is a mini science lesson.

The Motorboat Book

Build & Launch 20 Jet Boats, Paddle-Wheelers, Electric Submarines & More

Ed Sobey

978-1-61374-447-5
$14.95 (CAN $16.95)

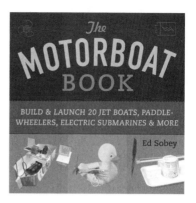

Ahoy, matey! There's more than one way to power a toy boat! Electric motors, balloons, gears, pumps, belt drives, chemical reactions, steam, fans, and even gravity can be used to propel a small ship across a pool. *The Motorboat Book* will show you how to build more than 20 different models using illustrated, step-by-step instructions. And if you'd rather travel under the water than over it, this book also has six different submarine projects. Most of the boats are built from recycled and easy-to-find materials, but an appendix lists local and online sources for where you can buy pulleys, plastic propellers, small motors, and more.

Unscrewed

Salvage and Reuse Motors, Gears, Switches, and More from Your Old Electronics

Ed Sobey

978-1-56976-604-0
$16.95 (CAN $18.95)

Unscrewed is the perfect resource for all UIYers—Undo It Yourselfers—looking to salvage hidden treasures or repurpose old junk. Author Ed Sobey will show you how to safely disassemble more than 50 devices, from laser printers to VCRs to radio-controlled cars. Each deconstruction project includes a "treasure cache" of the components to be found, a required tools list, and step-by-step instructions, with photos, on how to extract the working components. It also includes suggestions on how to repurpose your electronic finds. Fight the mindset of planned obsolescence—there's technological gold in that there junk!

The Way Toys Work

The Science Behind the Magic 8 Ball, Etch A Sketch, Boomerang, and More

Ed Sobey and Woody Sobey

978-1-55652-745-6
$14.95 (CAN $16.95)

"Perfect for collectors, for anyone daring enough to build homemade versions of these classic toys and even for casual browsers." —*Booklist*

Profiling 50 of the world's most popular playthings—including their history, trivia, and the technology involved—this guide uncovers the hidden science of toys. Discover how an Etch A Sketch writes on its gray screen, why a boomerang returns after it is thrown, and how an RC car responds to a remote control device. This entertaining and informative reference also features do-it-yourself experiments and tips on reverse engineering old toys to observe their interior mechanics, and even provides pointers on how to build your own toys using only recycled materials and a little ingenuity.

The Way Kitchens Work

The Science Behind the Microwave, Teflon Pan, Garbage Disposal, and More

Ed Sobey

978-1-56976-281-3
$14.95 (CAN $16.95)

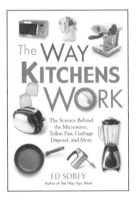

If you've ever wondered how a microwave heats food, why aluminum foil is shiny on one side and dull on the other, or whether it is better to use cold or hot water in a garbage disposal, now you'll have your answers. *The Way Kitchens Work* explains the technology, history, and trivia behind 55 common appliances and utensils, with patent blueprints and photos of the "guts" of each device. You'll also learn interesting side stories, such as how the waffle iron played a role in the success of Nike, and why socialite Josephine Cochran *really* invented the dishwasher in 1885.

Available at your favorite bookstore,
by calling (800) 888-4741, or at
www.chicagoreviewpress.com

CHICAGO
REVIEW
PRESS